D0500045

URBAN REVITALIZATION

Volume 18, URBAN AFFAIRS ANNUAL REVIEWS

INTERNATIONAL EDITORIAL ADVISORY BOARD

ROBERT R. ALFORD
University of California, Santa Cruz

HOWARD S. BECKER
Northwestern University

BRIAN J. L. BERRY
Harvard University

ASA BRIGGS
Worcester College, Oxford University

JOHN W. DYCKMAN
University of Southern California

T. J. DENIS FAIR
University of Witwatersrand

SPERIDIAO FAISSOL
Brazilian Institute of Geography

JEAN GOTTMANN
Oxford University

SCOTT GREER
University of Wisconsin, Milwaukee

BERTRAM M. GROSS
Hunter College, City University of New York

PETER HALL
University of Reading, England

ROBERT J. HAVIGHURST
University of Chicago

EHCHI ISOMURA
Tokyo University

ELISABETH LICHTENBERGER
University of Vienna

M. I. LOGAN
Monash University

WILLIAM C. LORING
Center for Disease Control, Atlanta

AKIN L. MABOGUNJE
University of Ibadan

MARTIN MEYERSON
University of Pennsylvania

EDUARDO NEIRA-ALVA
CEPAL, Mexico City

ELINOR OSTROM
Indiana University

HARVEY S. PERLOFF
University of California, Los Angeles

P.J.O. SELF
London School of Economics and Political Science

WILBUR R. THOMPSON
Wayne State University and
Northwestern University

URBAN
REVITALIZATION

Edited by
DONALD B. ROSENTHAL

Volume 18, URBAN AFFAIRS ANNUAL REVIEWS

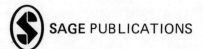 SAGE PUBLICATIONS   Beverly Hills   London

HT
175
U82
x

Copyright © 1980 by Sage Publications, Inc.

All rights reserved. No part of this book may be reproduced or utilized in any form or by any means, electronic or mechanical, including photocopying, recording, or by any information storage and retrieval system, without permission in writing from the publisher.

For information address:

SAGE Publications, Inc.  SAGE Publications Ltd
275 South Beverly Drive 28 Banner Street
Beverly Hills, California 90212 London EC1Y 8QE, England

Printed in the United States of America

Library of Congress Cataloging in Publication Data

Main entry under title:

Urban revitalization.

   (Urban affairs annual reviews ; v. 18)
   1. Urban renewal–United States.   2. Urban policy
–United States.   I. Rosenthal, Donald B., 1937-
II. Series.
HT108.U7 vol. 18 [HT175]   307.7'6s   79-27881
ISBN 0-8039-1190-4   [307.7'6'0973]
ISBN 0-8039-1191-2 pbk.

FIRST PRINTING

86377

CONTENTS

Preface

☐ PLANS FOR THIS VOLUME originated while five of the participants (Cho, Cicin-Sain, Gist, Rosenthal, and Stowe) were serving as Faculty Fellows of the National Association of Schools of Public Affairs and Administration (NASPAA) in the Department of Housing and Urban Development (HUD) during the academic year 1977-1978. Rosenfeld previously served there in the same capacity. Four other contributors have been or are currently associated with HUD (Abravanel, Mancini, Puryear, and Ross). Nevertheless, as readers of this volume will soon discover, these essays represent a wide range of views about appropriate policies and organizational mechanisms for promoting the redevelopment of American cities.

Because of its origins, the bias of the volume is toward a national perspective on urban revitalization. Aside from the Hula essay, which makes use of data on institutional lending practices in Dallas, we are dealing with national trends and national policy approaches rather than reporting the great variety of local differences that exist with respect to revitalization strategies. Even the Davies and Meehan chapters, which deal with materials from Great Britain and Latin America, respectively, do so from a national policy perspective. Still, we have attempted to identify and tried to deal in some detail with matters of considerable importance to anyone interested in an overview of local efforts to promote the revitalization of cities.

In the course of bringing this volume to completion, a number of persons have been especially helpful to me. I would like to express appreciation to Sanford Miller and Harry Wessel and, especially, to Lawrence Ribler, all of whom provided valuable research assistance. Most important of all, however, was the contribution of Betty Balcom, who was extraordinarily understanding of the tedious demands that I made during the lengthy process of preparing the final product for publication.

—Donald B. Rosenthal

Editor's Introduction

DONALD B. ROSENTHAL

☐ THE PRESENT VOLUME deals with selected aspects of the urban
revitalization process. It appears at a time when important changes may be
taking place in the images if not the reality of life in the central cities of
American metropolitan areas. A very few years ago, even guarded optimism about the future of central cities seemed out of place. Instead,
discussions focused on ways of coping with the restricted futures anticipated for city populations including strategies for "opening up the
suburbs" (Downs, 1973).

Current diagnoses of the fate of the city are no longer influenced solely
by images of decline and defeat. Although the factors that produced the
"urban crisis" are far from being excised, a new theme of hopefulness
seems to have entered into discussions of the state of the cities. Reinvestment in central business districts (CBDs) is evidenced by projects such as
Detroit's Renaissance Center, Boston's Quincy Market, and Baltimore's
waterfront and Charles Plaza developments—all of which have received
considerable favorable publicity. Equally important, the resurgence of
neighborhoods in cities across the country through processes to which
such labels as "gentrification" or "the back-to-the-cities movement" have
been attached calls attention to new investment by young couples and
single-person households and to the upgrading activities that have been
initiated by or on behalf of longer-term residents through the use of an
enhanced flow of private and governmental resources.

The shift in imagery may ultimately have beneficial consequences for
how we behave toward older central cities. Yet, as several of the chapters
in this collection indicate, such empirical evidence as we now have about

revitalization is fragmentary at best. It by no means supports a picture of a thoroughgoing comeback. Indeed, among the major purposes of this volume is not only to highlight some of the issues related to revitalizing our cities but to review the problems associated with measuring and analyzing those changes which have actually gone on.

Like beauty, the meaning of "urban revitalization" is very much in the eye of the beholder. For example, the *shape* of the future city is open to conjecture. At least three normative models may be found in discussions of the subject. First, one might seek to reconstitute the central city that existed prior to World War II. To realize such an extreme goal, of course, there would need to be a radical reversal of those economic and social relations which have largely been destroyed in the course of the past forty years. Even the energy crisis, which is seen as a major force likely to reverse decentralizing trends in the future, may have arrived too late to make more than a marginal impact on the shape of American metropolitan areas. Given not only the enormous movement of employment opportunities and middle-class populations to the suburbs but the character of shifts in the structure of the American economy itself, promotion of reinvestment in economic organizations of the kind that created the nineteenth-century city appears to be unrealistic.

A more realistic alternative implicit in the policies of the recent past involves a vision of a scaled-down but revitalized central city. Recent reinvestment in urban infrastructure and physical structures may capture for central cities some of the benefits which flow from shifts in the economy including the changeover from an industrial to a service and consumption economy (Harrison, 1974). The hallmark of this second model, however, is selective revitalization, which means focusing resources on those special features of each city most likely to attract and hold business and residential investment. Such a strategy may involve the partial abandonment of parts of older cities while concentrating investment in those areas where reinvestment is most likely to turn patterns of socio-economic organization around. This may mean, for instance, providing only limited attention to already healthy business and neighborhood districts.[1]

In any case, parts of central cities continue to display certain advantages of an institutional and economic character which make them attractive to a range of populations. In an age of diverse life-styles, high disposable incomes and small household units among the middle class, central city residence may be especially attractive for such groups. Thus, a third model of the urban future exists. Obviously related to the second, it suggests an even less ambitious set of goals. It might be characterized as "suburbanizing the central city" and depends upon making the central city attractive and "safe" again to the middle class. Such an image is consistent with the notion of multinodal metropolitan areas in which the central city becomes simply one of many specialized economic and social locations which prosper from peculiar advantages in terms of housing cost, locational advantages in a metropolitan area, and access to diverse educational, entertainment, and life-style options. Indeed, central cities may have already paid a great part of the heavy price levied upon them for the socioeconomic reorganization of metropolitan areas that took place after World War II. As Leven (1979) recently has suggested, in the future the great losers may be the second- and third-ring suburbs which have relatively less flexibility in recycling their metropolitan functions.

If there is possible disagreement about the future shape of the revitalized city, there is also likely disagreement about its *content*. For someone like Jacobs (1963), the joy and vitality of urban living lie in the proximity of social juxtapositions, diverse economic opportunities and "humane" but varied physical structures. For others, however, the life of a city is dependent on the strength of its "communities"—neighborhoods which may be ethnically, socially, and economically homogeneous and even mutually antagonistic while making up the larger urban mosaic (Long, 1972).[2] For still others, revitalization must center on building an autonomous economic base for those lower-income groups which remain the larger share of the central city population (Harrison, 1974).[3]

Whatever the goals for cities of the future and the different routes one might follow for getting there, it is important to understand where we are at the moment. The chapters in this volume deal both with the present condition and with various strategies for altering that condition. The first part in particular focuses on the former but also suggests some of the options available to individuals and to private interests operating in contemporary cities.

Thus, the chapter by Abravanel and Mancini highlights the fact that, in the aggregate, central cities continue to lose population. The main concern of that chapter is to show the limited prospects for a massive "back-to-the-cities" movement. The authors use a recent survey done for the Department of Housing and Urban Development (HUD) which identifies some of the attitudes which restrict optimism about prospects for revitalization. It must be understood, of course, that even the small percentage of respondents interested in moving to the cities from nonurban areas—as opposed to the much larger number of nonresidents who continue to value selected features of the central city—would have a significant impact upon urban life if their investments were directed toward a few cities or neighborhoods. We do not yet know how much that is the case.

Reinforcing the opinion data presented by Abravanel and Mancini, Cicin-Sain reviews what data exist with respect to actual reinvestment patterns in urban neighborhoods. She indicates that such investment has been both limited and highly selective. Her data suggest a need for caution about the prospects for a neighborhood revitalization "movement" beyond a few cities. Equally important, Cicin-Sain introduces a theme which will be heard elsewhere in the volume: urban revitalization has its costs as well as its benefits. In particular, she draws our attention to an issue which ran through the early years of criticism of urban renewal and threatened, at one time, to destroy that program: the social costs borne by those displaced by private reinvestment in central city redevelopment.[4]

Despite the uncertainties surrounding prospects for revitalization, private financial institutions have begun to support neighborhood reinvestment with some vigor in recent years. Richard Hula reviews some of the general issues and, more specifically, lending practices in the city of Dallas. His examination of "redlining" indicates just how complex that issue is even in terms of definition and measurement. Hula does find a consistent pattern of resource distribution in Dallas that favors some sections of the metropolitan area as opposed to others. It is disconcerting to learn, however, that what can and should be done about the situation in terms of public interventions is by no means clear. As useful as disclosure laws may be, and Hula's own research would have been impossible without the availability of data that resulted from such laws, the nature of the political system reduces prospects for bringing about major changes in the operation of the private investment market. Governmental intervention in this area, as Hula notes, may also be a two-edged sword, since reinvestment in currently redlined neighborhoods may simply accelerate displacement of lower-income groups.

It is not clear how much of the present interest in revitalization is a product of private investor choices and how much a consequence of recent government policies which have made a considerable amount of money available to central cities on a relatively unrestricted basis. While the Cicin-Sain and Hula chapters touch on aspects of the operation of the private market in neighborhood reinvestment—aspects which have enormous implications for public policy—the remaining contributions to the volume address more directly the role of government and the consequences for cities of national policies.

Prior to the recent efflorescence of federal grant programs to revitalize cities, the common wisdom had been that federal policies promoting major projects in metropolitan infrastructure (including highway) development and federal support of suburban mortgages accelerated, if they did not actually create, the process of mass suburbanization which sucked life from central cities. As a result, much blame has been laid at the doorstep of the federal government for the decline of the central city. No doubt, some of this blame is well deserved, although it largely reflected the way federal policy served the interests of the predominant economic and social forces in American society. Nonetheless, federal government leadership has been most notable in the past for its failures including its neglect in charting a course for urban revitalization. For the most part, federal policy-makers have generated a series of disconnected programs which have made resources available to local governments and to lower-income urban populations without anyone being in a position to consider the interactive effects of these programs. Thus, while programs like Model Cities attempted to bring together concern with both the physical and socioeconomic needs of deprived urban populations, the collapse of that program epitomized the failure of the national government both to develop a coherent national policy for the cities and to implement such a policy (Frieden and Kaplan, 1975).

Instead, the federal government has been much better at making resources available to local governments which the latter can target toward the realization of local purposes, particularly purposes which for the most part reflect the preferences of local interests groups for construction activities. In many ways, the original urban renewal program of the 1950s most clearly pursued this physical approach to revitalization. Under it, local plans for urban redevelopment wiped out neighborhoods and small businesses in the name of inducing middle-class populations either to remain in the city or to return. There was some hope that such an approach would not only strengthen the city's economic base but eventually benefit the poor, albeit through a "trickle-down" process. There has been much criticism of this strategy and, yet, as Heywood Sanders points out, the effects of urban renewal both in terms of what it removed from the urban landscape and what it actually added are still only partially understood. Many of the complaints which were justifiably directed at urban renewal during its early years (Bellush and Hausknecht, 1967; Wilson, 1966) require reexamination in light of changes in the design and operation of the program under the Johnson Administration. While some readers may disagree with Sanders's reinterpretation of events, his chapter presents a useful review of changes in the program over time which shifted program emphases away from commercial investment to housing and from clearance to rehabilitation.[5] Indeed, before the urban renewal program was melded into the Community Development Block Grant (CDBG) in 1974, as Sanders indicates, many of the central city projects which now are being praised for contributing to urban revitalization were just being moved into implementation.

Whatever the lessons to be drawn from the later history of the urban renewal program, it was never effectively geared toward concentrating resources on problems of the poor. The 1960s did, of course, bring forward other programs which responded more directly to what were perceived to be the needs of low-income groups. Whether attributable to urban unrest, to the growth of upper-middle-class sensitivities about the problems of the poor (Banfield, 1970), or, as Piven and Cloward (1971) have suggested, to the political recognition that urban blacks constituted a major political constituency available for mobilization by the Democratic party, programs like those incorporated in the War on Poverty, Model Cities, and the various forerunners of the Comprehensive Employment and Training Act (CETA) represented important federal initiatives bringing resources to low- and moderate-income citizens. Yet, these uses of federal resources involved payoffs essentially to nonbeneficiaries of American society's major contests for economic and social opportunity (Long, 1972; Sternlieb, 1971). The direct effects of such programs for the revitalization of central cities have to be judged as minimal.

In contrast, the aggregate effects of the enormous growth in the last decade of intergovernmental grants-in-aid may have made some contribution to the general welfare of cities as organizers of physical relationships, allowing their local governments the breathing space necessary to undertake major renovations of urban infrastructures and providing them with vital resources to stimulate CBD and neighborhood reinvestment. John Ross draws upon data from various programs—general revenue sharing (GRS), CDBG, CETA, the public works programs of the Economic Development Administration (EDA), and countercyclical aid—to consider how such aid impacted upon America's largest cities during the period from 1960-1977. In the course of his review, he raises important questions about changes in the political roles local government actors have come to play in intergovernmental relations. Thus, while the growth in aid has been dramatic, it is by no means clear that federal *control* over the design and implementation of local efforts to use those funds on specific projects has increased. If anything, local power may well have been enhanced as block grants replaced earlier project grants. They have allowed many cities to respond to neighborhood demands of the kind that received only limited encouragement during the heyday of urban renewal. At the same time, however, substantial sums from CDBG have also gone into closing out urban renewal project in CBDs and furthering other CBD projects.

While many of these grant programs were products of the Nixon-Ford years, the Carter Administration has pursued a course which has sought to change the direction of the New Federalism only at the margins, not to reassert the national leadership style characteristic of the Johnson years. Terminologically, the approach which has emerged emphasizes a "partnership" among governmental actors, including state governments, and with the private sector and neighborhood groups.

In line with his managerial style, however, President Carter has attempted to reshape those aspects of federal programs which tended to promote the inefficient application of federal resources to urban problems. At one level, this has resulted in a series of efforts to pinpoint—*target*—program resources to those cities and neighborhoods most in need. At another level, it has meant a conscious effort to formulate a national approach to problems of urban policy. The two chapters that close out the second part of the volume focus on the second approach.

As Eric Stowe illustrates, the process of formulating a national urban policy was beset by political jockeying among bureaucratic actors. Stowe, who was a close observer of the process, does not share with Anthony Downs (1978) the view that the behavior of participants in the urban policy-making process followed a rational course of development which first identified the goals of such a policy and then moved on to determine the appropriate means for realizing those goals. Instead, Stowe suggests that the bureaucratic quarrels which almost capsized the process at several points resulted in an outcome which appears to have consigned the final recommendations of the report to the scrap heap.

While the effects of the programmatic efforts of the 1960s may be debated, there is considerable evidence that the period left in its wake a congeries of neighborhood organizations. Some of these were stimulated originally by resistance to federal projects (Fainstein and Fainstein, 1974); others were given an impetus by the availability of federal funds. Janice Perlman (1976; 1979) has shown how such groups have come to play an important part in contemporary revitalization decisions quite different from the grudging recognition of early urban renewal efforts. Although many of the individual neighborhood groups have limited goals and tend to be largely absorbed in their distinct local situations, these local bodies have also taken some tentative steps toward making themselves felt in the national policy process (New York Times, 1979g.)

The Carter presidential campaign of 1976 responded to the phenomenon of the mobilized neighborhood in its appeals to both ethnic and minority identifications. Subsequently, HUD recognized the "neighborhood factor" by creating a new Office of Neighborhoods, Voluntary Associations and Consumer Protection which provides a vehicle for local groups to be heard within the councils of national urban policy-makers.

Even before Carter took office, however, the Ford Administration and Congress supported the formation of a National Commission on Neighborhoods, which was timed to go into operation after the elections of 1976. The commission membership, which was actually appointed by the new Carter Administration under rules set earlier, responded to this mixed parentage by seeking to incorporate a variety of political and ideological positions. As in the case of the national urban policy described by Stowe, the course of producing a report by the Neighborhood Commission was far from smooth. As the chapter by Barry reveals, representatives of working-class ethnic neighborhoods and spokespersons for low-income minority neighborhoods held different perspectives. More important, however, the internal organization of the commission was dominated by a small group of commissioners who exercised quite diverse influences over the work of the commission with the result that it ultimately produced a voluminous report which attempted to paper over disagreements by offering something for everyone. Barry, who worked as a staff member of the commission during its last days, ends on a hopeful note, but one which owes less to the probable impact of the commission's recommendations than to its identification of major issues in revitalization which are likely to be the subject of much attention in the 1980s.

As noted earlier, one aspect of the Carter Administration's efforts has been to reinforce a targeting approach to resource distribution. The literature on intergovernmental relations suggests that the American political system is much better equipped to follow a *distributive* approach to resource allocations—spreading federal resources widely but thinly—rather than following a *redistributive* approach which concentrates national resources in a massive fashion on areas where the need is greatest (Lowi, 1964; Beer, 1973). Nevertheless, the contributions to the third part of the volume highlight programmatic efforts to focus national resources on those localities and populations most in need of assistance.

The paper by Cho and Puryear explores the targeting effort by looking at both alternative formulas for delivering resources to cities and the distributive effects of various HUD programs. On the whole, they find that while the particular formula does not make much difference, HUD has been reasonably successful in its pursuit of a targeted approach. Some formula-based programs are clearly more distributive in their impacts than those programs which utilize project-grant approaches, but that is to be expected given the design of the particular project grants.

In a similar vein, Rosenfeld reviews the controversy within HUD and with Congress over assuring that a significant proportion of CDBG funds is aimed toward meeting the needs of low- and moderate-income urban populations within metropolitan areas. He suggests in part how difficult implementing that goal has been even in terms of sorting out the definitional and measurement problems. Interestingly enough, even with the crude measures available, Rosenfeld finds some evidence that the critics of CDBG, and of the New Federalism generally, may be incorrect when they claim that the program has diverted local governments from the use of resources in ways that favor low- and moderate-income populations. Indeed, the evidence he provides suggests that such populations may have done better under the uncertain standards that prevailed before Carter came into office than under the tighter definitions recently promulgated by Congress and the Administration. As Rosenfeld illustrates, the Carter Administration attempted at least until recently through regulatory controls to tighten the obligations of local governments to meet the needs of the poor. Whether that approach has backfired politically remains to be seen.

At the same time, the Administration sought to develop new programs which would encourage middle-class and business reinvestment in central cities. Indeed, economic development was a central theme of the ill-fated national urban policy as evidenced in the various recommendations for a National Development Bank and the continuous references to cooperation between the private and public sectors. It is too early to undertake analyses of the preliminary impacts of the newer programs, but Gist describes one such initiative by HUD to stimulate urban reinvestment through the Urban Development Action Grant (UDAG) program. Gist notes how UDAG revived certain aspects of the urban renewal approach. In the spirit of the Administration's concern with targeting, however, complex procedures were adopted by HUD, first to identify those localities most in need of UDAG program resources which would qualify them for application and subsequently to select from among the qualified those projects which met HUD requirements for funding. On the whole, Gist finds that initial qualification and ultimate selection were rather distinct processes with the result that the final choice of projects was by no means a function of the ability of a project to stimulate reinvestment only in the neediest cases.

The distributive approach emphasized in congressional attitudes to resource distribution also has recently moved UDAG further from this effort at targeting. In the same way that Rosenfeld reports Congress has exercised pressure on HUD to loosen its commitment to the targeting of CDBG funds to low- and moderate-income central city populations, representatives of the South and Southwest have recently succeeded in including certain prosperous cities in the UDAG program by qualifying those with "pockets of poverty" for participation (New York Times, 1979e, 1979f).[6]

While some UDAG projects repeat the emphasis of early urban renewal efforts on commercial investment, the growth in influence of neighborhood groups has placed an important constraint on the program not present in the first phase of the earlier program. In particular, such groups have demanded the location of UDAG projects in neighborhoods rather than exclusively in CBDs and have received concessions, although, as Gist indicates, some question may be raised about the labeling of certain UDAG projects as "neighborhood" in character.

In 1978, the Carter Administration introduced another program to stimulate revitalization of neighborhoods, the Neighborhood Strategy Area (NSA) program. That program allocates rental subsidies from the Substantial Rehabilitation portion of the Section 8 program to those neighborhoods selected by local governments as target areas for concentrated investment of public and private resources. In contrast to earlier HUD programs, which assigned federal housing subsidies to specific sites largely in response to the preferences of private developers, it was now the responsibility of the local government to coordinate the variety of activities necessary for the revitalization of a given neighborhood: improvement of physical facilities; provision of necessary public services; assurance of access to adequate commercial and personal services; rehabilitation not only of rental units but of owner-occupied residential structures. Thus, as with the UDAG program, NSA attempted to leverage limited federal program resources in a manner that would trigger the expenditure of substantial amounts from a wide range of other government programs and from private sources so as to realize significant housing and community improvements within the designated neighborhood. In September 1978, HUD selected 155 neighborhoods in 118 cities (from approximately 211 neighborhoods for which applications were submitted) for participation in the program. However, that was only the first of many steps in a complex implementation process which found most participants in late 1979 still at work putting together the financial packages and developing site plans for rehabilitation of specific buildings within their NSAs. (For a detailed review of the emergence of the program, see Rosenthal, 1979.)

Finally, the chapters by Davies and Meehan provide us with the basis for comparisons with other societies. Thus, the Davies chapter examines the British approach to urban revitalization and suggests many parallels to the American experience. From programs of clearance and CBD redevelopment, the British have gradually moved to accept the notion that rehabilitation of existing housing and serious attention to the total environment of an urban neighborhood are important parts of any national urban policy. Unfortunately, this new approach has been slow to mature and has run into difficulties both in design and implementation. As in the United States, the commitment of national resources to implement local program designs may prove to be the ultimate weakness of this effort. As the

British economy has moved through recession and inflation and back again, the ability and willingness of the national government to provide the large resources necessary to rehabilitate neighborhoods in a planned fashion have already run into difficulties. That may well prove to be a major problem in the United States as well. It remains to be seen, for example, how successful the NSA effort will be if the recession of 1979-1980 and the budgetary control measures which are currently being put in place operate as expected. It is discouraging, for example, that a major cutback in housing subsidies already seems to be underway (New York Times, 1979d).

Finally, Meehan's chapter reviews some of the sources of failure in American approaches to the problems of urban revitalization. Drawing upon lessons learned from self-help experiments in Latin America, he emphasizes the need to place the development of human capital at the heart of any strategy for revitalization. The particular projects he describes may be inappropriate for transfer to the United States, but his chapter highlights an important theme touched upon by others in this volume only in passing: government investment in physical renewal projects without adequate investment in meaningful opportunities to develop economic skills among the urban poor is likely to limit our capacity to make American cities truly livable.

As noted at the outset of this essay, this volume arrives at a moment when a concern with the "urban crisis" is being muted through a new emphasis on urban hopefulness. How much this hopefulness can be attributed to the activities of the Carter Administration is open to debate. Nonetheless, the volume also represents a summing up of some of the efforts undertaken during the years of that Administration to stimulate urban reinvestment. It is difficult to abstract any single theme or set of themes with which to summarize government approaches to urban revitalization during the Carter years. On the one hand, the emphasis on targeting suggests a sense of limited growth in urban aid. Programs like UDAG and NSA reflect an effort to target federal resources in a highly selective fashion. At the same time, however, HUD appeared ready at one point to make a major commitment in response to New York City's demand for billions of dollars in federal funds toward the revitalization of the South Bronx--a commitment that grew out of the President's highly publicized visit to New York in October 1977. These plans now appear to have fallen through (New York Times, 1979b), presumably much to the relief of the Administration. HUD leadership was also forced to bend to political considerations in recognizing the claims of some doubtful candidates for revitalization under the NSA program and in accepting the qualification of less needy cities for participation in the UDAG program. Yet, the Administration continues to verbalize support for targeting federal investment on those urban opportunities which make the greatest economic sense rather than investing in grandiose plans which threaten to become bottomless pits for nonproductive investments, albeit in the name of a concern for supporting the poorer sections of urban society.

It is very difficult politically to follow a course which explicitly commits government to a policy of triage. Yet, there is nothing inherently wrong with choices involving the abandonment of neighborhoods which have already reached a point where crime is rife, the costs of maintenance are prohibitive and proposals for rehabilitation are so expensive as to be unrealizable. Of course, to conduct a policy of humane relocation and to provide alternative housing opportunities may also involve great costs. As a result, federal and local policies are likely to continue to support programs which waste considerable resources on cosmetic improvements with little potential for long-term revitalization.

Where this leaves prospects for urban revitalization as a whole is impossible to predict. As virtually every contribution to this collection suggests, there are innumerable difficulties in the way of implementing and evaluating revitalization-related programs. Indeed, because we are so early in the process, we do not yet have either an adequate taxonomy of revitalization or an understanding of how it differs from one locality to another. Indeed, one of the more obvious shortcomings of this volume is that so much of the emphasis is on the general and national rather than the particular and local. Disaggregation might show, as Nathan has suggested, that some urban regional centers have excellent prospects for revitalization while others do not (New York Times, 1979a). The former cities may continue to serve as major regional centers and to attract new commercial and residential investment that may well prove to be self-sustaining. Boston, Chicago, perhaps Baltimore, may warrant such confidence. There is no question that metropolitan areas like New York, Washington, and San Francisco will continue to perform many important functions as national and international centers. What is less clear is whether other large cities—Cleveland, St. Louis, Newark, Buffalo—have the same potential for economic and residential revitalization. Of equal interest are the many small cities which once fulfilled important subregional economic roles but whose futures look bleak. For many of these small cities of the Northeast and Midwest, the concept of "urban revitalization" may amount to little more than investment in symbolic projects unrelated to prospective economic vitality. Federal resources which support projects in such places may be desirable in the same way that the hospice may be preferable to the terminal ward, but they can hardly be characterized as rational investment of limited national resources nor do they exemplify realistic national urban revitalization strategies.

NOTES

1. In that respect, see the political controversy generated by the politically imprudent (albeit realistic) suggestions of Roger Starr for disinvestment in certain neighborhoods—the adoption of a strategy of "triage" for cities—while he was serving as Administrator of Housing and Development in New York City (New York Times, 1976).

2. In the first case, we might expect policies which would encourage middle-class investment so long as displacement is minimized. This assumes acceptance by the middle class of lower-class neighbors once they have established a significant beach-head—an arrangement which has been rare except in neighborhoods of the kind Jacobs describes for the Greenwich Village area. The second type of neighborhood, more characteristic of current discussions of "defensible space," would *discourage* diversity through policies which would limit opportunities for racially integrated housing or school busing. The socioeconomic diversity of a neighborhood does not mean, of course, that social interaction actually occurs. As Suttles (1972) has shown, persons of diverse ethnic backgrounds may live next to each other but move in different social spaces. That may well have been the case even in Jacob's Manhattan neighborhoods. For a useful discussion of different themes within the "neighborhood" literature, see, Stever, 1978.

3. After an early flurry of enthusiasm for community development corporations as vehicles for the stimulation of economic development, there appears to have been a general decline in their level of economic activity. The best-known, such as the Bedford-Stuyvesant Corporation, on closer examination tend to be run very much along the lines of private businesses, often being dominated by individual entrepreneurs. Others founded in the 1960s have either gone out of operation or linger on largely as vehicles for local participation in the delivery of services unrelated to economic development or as pass-through agencies for federal grants. On the original hopes, see Kotler, 1969, and Harrison, 1974. For a more recent evaluation, see Garn et al., 1976.

4. Because of the rapidity with which the "gentrification" process was working in Washington, DC, at the time the Carter Administration came into office, special concern about the displacement issue arose there. Nevertheless, a congressionally mandated study of the issue reported relatively little cause for alarm (New York Times, 1979c). On displacement, also see Gale, 1978; Kollias, 1978; and U.S. Senate, 1977.

5. For a study of urban renewal efforts in Boston that reflects the transitional phase that Sanders discusses, see Keyes, 1969.

6. Unfortunately, none of the chapters deals directly with issues related to the responsibility of states or suburban governments, for that matter) for the current condition of their central cities. From one perspective, it should not be necessary for federal funds to be used to aid those central cities whose immediate suburbs or state governments are capable of supporting the urban revitalization process. Except for a few states like Massachusetts and New York, however, there has been a limited commitment by state governments to urban reinvestment. Under the circumstances, demands by cities like Houston or Phoenix that they participate in the UDAG program rather than providing their own resources to their poorer neighborhoods are consistent with the general attitudes fostered by the American style of distributive politics.

REFERENCES

BANFIELD, E.C. (1970) The Unheavenly City. Boston: Little, Brown.

BEER, S.H. (1973) "The modernization of American federalism." Publius 3 (Fall): 49-95.

BELLUSH, J. and M. HAUSKNECHT (1967) Urban Renewal. Garden City, NY: Doubleday.

DOWNS, A. (1973) Opening up the Suburbs. New Haven, CT: Yale Univ. Press.

——— (1978) "Urban policy," pp. 161-194 in J. Pechman (ed.) Setting National Priorities: The 1979 Budget. Washington, DC: Brookings.

FAINSTEIN, N.I. and S.S. FAINSTEIN (1974) Urban Political Movements. Englewood Cliffs, NJ: Prentice-Hall.

FRIEDEN, B.J. and M. KAPLAN (1975) The Politics of Neglect. Cambridge, MA: MIT.

GALE, D.E. (1978) "Dislocation of residents endangers neighborhood conservation efforts; what solutions are possible?" Journal of Housing 35 (May): 232-235.

GARN, H.A., N.L. TREVIS, and C.E. SNEAD (1976) Evaluating Community Development Corporations—A Summary Report. Washington, DC: Urban Institute.

HARRISON, B. (1974) Urban Economic Development. Washington, DC: Urban Institute.

JACOBS, J. (1963) The Death and Life of Great American Cities. New York: Vintage.

KEYES, L.C. (1969) The Rehabilitation Planning Game. Cambridge, MA: MIT.

KOLLIAS, K. (1978) "Revitalization without displacement," HUD Challenge 9 (March): 6-7.

KOTLER, M. (1969) Neighborhood Government. Indianapolis: Bobbs-Merrill.

LEVEN, C. (1979) "Economic maturity and the metropolis' evolving physical form," pp. 21-44 in G.A. Tobin (ed.) The Changing Structure of the City: What Happened to the Urban Crisis. Beverly Hills: Sage Publications.

LONG, N. (1972) The Unwalled City. New York: Basic Books.

LOWI, T. (1964) "American business, public policy, case studies and political theory." World Politics 16 (July): 677-715.

New York Times (1976) "City housing administrator proposes 'planned shrinkage' of some slums." (February 3).

——— (1979a) "Cities in poor shape getting worse but others improve, a study finds." (January 21).

——— (1979b) "South Bronx plan voted down 7 to 4 by estimate board"; "White House aides stunned by action." (February 2).

——— (1979c) "US housing study finds displacement of poor in slums is minimal." (February 13).

——— (1979d) "House pares subsidized housing." (June 8).

——— (1979e) "Urban development grant: spur to private investment." (June 8).

——— (1979f) "With new political mood, focus of urban aid may ignore the poor." (June 11).

——— (1979g) "Activist neighborhood groups are becoming a new political force." (June 14).

PERLMAN, J. (1976) "Grassrooting the system." Social Policy 7 (September-October): 4-20.

——— (1979) "The neighborhood movement in the 1970s: grassroots empowerment and government response." Mimeo (April).

PIVEN, F. and R. CLOWARD (1971) Regulating the Poor. New York: Vintage.

ROSENTHAL, D.B. (1979) "Joining housing rehabilitation to neighborhood revitalization: the neighborhood strategy area program." Presented 1979 annual meeting of the American Society for Public Administration, Baltimore, April 1-4.

STERNLIEB, G. (1971) "The city as sandbox." Public Interest 25 (Fall): 14-21.

STEVER, J.A. (1978) "Contemporary neighborhood theories: integration v. romance and reaction." Urban Affairs Quarterly 13 (March): 263-284.

SUTTLES, G. (1972) The Social Construction of Communities. Chicago: Univ. of Chicago Press.

U.S. Senate (1977) Committee on Banking, Housing and Urban Affairs. Neighborhood diversity: Hearings on Problems of Dislocation and Diversity in Communities Undergoing Revitalization Activity. 95th Congress, 1st session.

WILSON, J.Q. (1966) Urban Renewal: The Record and the Controversy. Cambridge, MA: MIT.

PART I

RESOURCES FOR REVITALIZATION
The Nongovernmental Sector

1

Attitudinal and
Demographic Constraints

MARTIN D. ABRAVANEL and PAUL K. MANCINI

☐ MANY OF AMERICA'S LARGER URBAN CENTERS have been undergoing a process of decline characterized by steady population losses, high amounts of unemployment, increasing concentrations of poverty, severe fiscal stress, environmental pollution, and neighborhood and housing deterioration (U.S. Department of Housing and Urban Development, 1978b). Recently, however, some observers have begun to report signs of a reversal in this trend. For example, T. D. Allman (1978), in his article entitled "The Urban Crisis Leaves Town," argues that the urban crisis is waning, a renaissance beginning, and the cycle of distress is reversing. If a reversal has, indeed, begun to take place, this would have an obvious bearing on the kinds of strategies that the public sector might employ to encourage the growth and vitality of urban America. The following questions, therefore, are critical. Is the city now more attractive to many of the types of persons who have, for the last decade or so, been leaving it for the suburbs? Has the general public, or specific segments of it, begun to perceive the city as a desirable residential environment?

To understand better the attitudinal context for urban revitalization efforts, this chapter will examine data from a recent extensive survey on

AUTHOR'S NOTE: *The data reviewed here were drawn from a survey conducted under contract to HUD's Office of Policy Development and Research by Louis Harris and Associates. The findings and conclusions presented in this article are those of the authors, and neither Louis Harris and Associates nor HUD bear any responsibility for the analysis or interpretations of the data included in this article.*

public attitudes toward the quality of community life. It will focus on popular attitudes about cities and, more specifically, on the views and concerns of people who have recently moved or who contemplate moving to or from the nation's central cities. Of course, many factors are involved in the processes of urban decay or urban revitalization, and not all of them are based on public preferences or perceptions. There is little question, however, that public attitudes place an important constraint on these processes and that they place limits on the extent to which community or economic development programs can achieve their intended objectives.

SIGNS OF URBAN DISTRESS

Between 1960 and 1970, central cities lost 345,000 people per year through net migration. From 1970 to 1975, 13 million persons moved out of central cities and 6 million persons moved into them, resulting in a net loss of 7 million residents (Peterson, 1976). When the three major components of population change—natural changes caused by births and deaths, and international and interregional migration—are taken into account, central cities experienced an absolute population loss of almost 2 million people (or 3.1% of their total population) from 1970 to 1975 (Berry and Dahmann, 1977). From 1975 to 1977, 1,018,000 more people moved from central cities than moved into them. Hence, the total population of central cities declined by 4.6% between 1970 and 1977. During the same years, suburban areas experienced a 12% increase in total population.

For the nation's larger central cities, this population loss was even greater. For example, the population of central cities in metropolitan areas of one million or more persons declined by 7.1% during this seven-year period. In terms of numbers of persons, then, central cities have steadily declined and no net reversal of this trend is yet evident.

Although central cities have been losing population, the number of central city households has actually increased by 1.5 million or 7.2% during this period. On the surface, this type of growth may appear to be a positive indicator of the vitality of cities, but compared to the rate of household growth for suburban areas (22.5%) and nonmetropolitan areas (20.3%), central cities are clearly lagging. Furthermore, it has been argued that increases in the number of households do not appear to be the result of central cities attracting new households from noncentral city areas. Instead, larger families are moving out of central cities, smaller families and single individuals are splitting off from family units and, therefore, those remaining comprise smaller household units (Yentis, 1978).

The decline in central city population has been economically devastating, as higher-income and employed persons have been leading the exodus. The income levels of in-migrants have, at the same time, been lower than those of out-migrants. Because of declining populations and income differentials between in- and out-migrants, central cities experienced a net loss of $29.6 billion in aggregate personal income between 1970 and 1974 and an additional $17 billion between 1975 and 1977. The fiscal pressures on central cities that have resulted from these losses in family income were even further exacerbated by the job losses suffered by central cities during this period. Between 1970 and 1977, employment of central city residents declined by 1.4% while employment of suburban and rural residents increased by 25.7%.

The nation's central cities now contain a large and increasing proportion of the country's dependent populations, and this has contributed to the economic decline and fiscal dilemma faced by many of them. George Peterson (1976: 44), for example, has concluded that "cities that are losing population actually spend more, per capita, in capital investment than cities that are gaining population, due largely to the necessity and difficulty of replacing their antiquated capital stock." The need for and costs of police and fire services, for example, have not declined in proportion to population decline. In a recent study of the fiscal condition of cities (U.S. Congress Joint Economic Committee, 1977) it was found that, compared to other cities, those with high unemployment and decreasing populations exhibit the most symptoms of need and fiscal strain. These types of cities were forced to decrease their aggregate service and capital budgets between 1976 and 1977. They also experienced the largest aggregate reduction in unencumbered surplus and the largest total deficit.

SIGNS OF URBAN RENAISSANCE

Despite these bleak demographic, social, and economic trends, some observers have argued that cities are beginning to enter a period of renaissance. Thus a recent *New York Times Magazine* article (Fleetwood, 1979), argued that New York City may be entering a period of revival because the people who are moving into the city are young, professionals—lawyers, architects, doctors, and investment executives. The author concluded that "the evidence of the late 70s suggests that the New York of the 80s and 90s will no longer be a magnet for the poor and the homeless, but a city primarily for the ambitious and educated—an urban elite." By focusing on certain recent economic trends and drawing upon numerous anecdotes, T. D. Allman has maintained that many so-called distressed cities are no longer facing serious demographic, fiscal, economic, or social problems. Both he and the National Development Council in a recent publication (cited in Congressional Quarterly, 1979) list many of the following hypotheses and economic forces as support for the argument that cities may be entering a new period of residential and commercial revival:

(1) The per capita income of those families remaining in cities will increase, while the population of these cities will continue to decline;

(2) the cities are attracting substantial amounts of foreign investment;

(3) younger, highly educated, upwardly mobile families and individuals are moving to the cities;

(4) the flow of federal funds to cities "has become a blizzard";

(5) cities are generating new employment faster than, or at least as fast as, many suburban areas;

(6) suburbs are largely filled, with little space remaining for expansion;

(7) many of the social problems that are normally associated with cities have also moved to the suburbs;

(8) increased energy costs have led many people to question the costs of suburban transportation and the benefits of suburban home ownership; and

(9) the cost of industrial land in cities is now lower than that of comparable suburban space.

The publication of these ideas has sparked an interesting and spirited debate as to whether an urban renaissance is really under way. In response to the above articles, for example, the U.S. Department of Housing and Urban Development (HUD) published two working papers that concluded that many central cities still face extremely severe economic, fiscal, social, and economic difficulties. It was argued that revitalization is largely confined to small areas in a limited number of central cities, disinvestment in cities still surpasses investment, poverty in cities continues to grow, the urban fiscal crisis is not over, and the back-to-the-city movement is overwhelmed by the continuing exodus from central cities. Two other recent HUD studies, based on analyses of Annual Housing Survey data, also concluded that the "existence of a sustained 'back-to-the-city' movement is challenged by these data" (Nelson, 1979: 8) and "the facts, it seems, aren't relevant" to those who insist that such a movement exists" (Yentis, 1978: 31). Finally, an examination of the changing condition of urban America by Richard Nathan and James Fossett (1978) suggests that the urban renaissance theory has been carried too far by its proponents and that "the most severely distressed cities do not show signs of improvement; quite to the contrary, their problems appear to have deepened and worsened."

DISTRESS OR RENAISSANCE?: THE PUBLIC VIEW

Part of the debate over whether urban America is continuing on the path of decay or is beginning on a path toward revitalization centers on the perceptions and preferences of the general public. How are cities viewed and what are their positive and negative attributes? What kinds of people have moved to and from the city and is there any evidence of changes in locational preferences? Why do people move to and from the city? What are the satisfactions and problems experienced by urban residents? While answers to these questions will not resolve the debate over the direction that urban America is taking, they will shed some light on the extent and possibilities for urban revitalization efforts in the near future.

To address the questions listed above, we will rely on data from a 1978 HUD survey on the quality of community life—a national, cross-section sample of 7,074 Americans, 18 years of age and older, conducted during November and December 1977.[1] The sample consisted of three primary strata:

(1) cities—every place defined as a central city of a Standard Metropolitan Statistical Area (SMSA) by the Bureau of the Census;
(2) suburbs—every place that is not a central city but is within an SMSA; and
(3) nonmetropolitan areas—cities, towns or villages and rural areas outside of SMSAs.

In-person interviews with respondents covered a variety of topics including community and neighborhood assessments, preferences, and satisfactions.

POPULAR ATTITUDES ABOUT CITIES

Although most Americans do not view cities as the most desirable places in which to reside, cities have not been written off as having no meaningful function to play. Nevertheless, as perceived by the citizenry, the city represents some of the best and some of the worst in American life.

The Negative Side of City Life

As surveys conducted over the last several decades have shown, most Americans believe that the best place to live is outside of urban areas. Less than a majority (41%) of current central city residents rate large cities or nonsuburban medium sized cities as the "the best place to live." An even smaller percentage of suburban residents (11%) and persons residing in nonmetropolitan areas (10%) see the city as the best residential environment (see Table 1.1). The belief by city residents that life is better outside of larger cities is partially a reflection of the perceived positive attractions of the countryside and partially a reflection of the perceived severity of the problems of urban communities.

More than anything else, the American public characterizes its large cities as having more crime and lawlessness than any other type of community. In addition, large majorities see cities as having the worst housing, the worst schools, and as being the worst places in which to raise children (see Table 1.2). The general image of cities as the reservoir of the country's social ills is underscored by data on the attitudes of city residents themselves toward their communities. When city residents were asked to rate the severity of fourteen potential community problems, crime was by far the most frequently cited "severe" problem (72%), followed by drug addiction (60%) and unemployment (60%). With the exception of a "lack of interesting things to do" and a "lack of parks," city residents rated all of the other problems as "severe" more frequently than did those who lived outside of central cities in evaluating their locales.

The view that problems are severe is also related to the objective condition of the community. For example, people who live in the nation's larger cities (250,000 or more) and in cities whose populations have

TABLE 1.1   Americans' Beliefs About the "Best Place to Live" by Place of Residence (percentages)

| | | Current Place of Residence | | |
Best Place to Live	National	Central City	SMSA Outside of Central City	Small Town/ Rural Area
Large city	10	26	3	3
Medium city not in the suburbs	10	15	8	7
Medium city in the suburbs	14	17	21	3
Small city, town or village in the suburbs	16	9	23	13
Small city, town or village not in the suburbs	20	10	18	31
Rural Area	24	12	22	39
It makes not difference	1	2	2	1
Not sure/don't know	5	9	3	3
Total	100	100	100	100
Number of Respondents	(7,074)	(3,298)	(3,229)	(547)

TABLE 1.2   Americans' Beliefs About the Attributes of Cities, Suburbs, and Rural Areas (percentages; N = 7,074)

Which Type of Community Has	Cities	Suburbs	Small Towns/ Rural Areas	There is No Difference/ Not Sure	Total
(A) Negative Attributes					
The most crime	92	1	1	6	100
The worst place to raise children	83	2	4	11	100
The worst housing	64	4	14	18	100
The worst public schools	63	3	15	19	100
The highest divorce rate	58	8	1	33	100
The highest taxes	58	19	4	19	100
The greatest amount of racial discrimination	45	14	18	23	100
The worst shopping facilities	14	6	66	14	100
(B) Positive Attributes					
The most plays, museums, and cultural opportunities	90	4	–	6	100
The best public transportation	81	8	2	9	100
The best selection of movie theaters	78	12	1	9	100
The best selection of restaurants	77	14	2	7	100
The best clinics, hospitals, health care facilities	73	15	4	8	100
The best employment opportunities	72	13	3	12	100
The best colleges and universities	67	13	4	16	100
The best shopping facilities	47	32	9	12	100
The best public services (garbage collection, fire and police protection, etc.)	39	33	15	13	100
The widest range of housing that you can afford	32	27	23	18	100
The least amount of racial discrimination	29	12	35	24	100
The best public schools	23	39	24	14	100
The greatest number of people who have attitudes similar to yours	23	24	33	20	100

TABLE 1.2   Americans' Beliefs About the Attributes of Cities, Suburbs, and Rural Areas (percentages; N = 7,074) (Cont)

Which Type of Community Has	Cities	Suburbs	Small Towns/ Rural Areas	There is No Difference/ Not Sure	Total
The best housing	22	43	18	17	100
The friendliest people	13	21	49	17	100
The lowest cost of living	13	10	54	23	100
The lowest taxes	12	11	53	24	100
The best place to raise children	12	30	49	9	100
The least Crime	4	14	69	13	100

declined in recent years were even more likely to see their communities as having social, economic, and physical problems than were people who reside in other types of places. In stark contrast, none of the fourteen problems was considered to be "severe" by more than 23% of residents of the noncentral city portion of SMSAs. Drug addiction ranked first in these suburban areas (23%), followed closely by traffic congestion (22%). Crime, ranked below traffic congestion, was mentioned by 20% of suburban residents.

The Positive Side of City Life

Despite the negative image of the city as a place to live among both the general public and city residents, and despite the widespread belief that life is better outside of urban settings, a large majority of those who live in central cities (71%) expressed satisfaction with the overall condition of their community (see Table 1.3). Close proximity to not only the worst, but also the best attributes of cities may help to explain this apparent inconsistency: central city residents, like others, also see a positive side to city life.

Cities, more than any other type of locale, are perceived by most Americans to provide the best shopping facilities, transportation services, and employment opportunities. They are also seen as having more plays, museums, movie theatres, restaurants, and universities. When asked for their recollections about the past, for example, Americans remember cities as having the best employment opportunities for both skilled and unskilled workers, and when asked about the present, this still appears to be the case: a clear majority—77% of central city residents, 70% of suburban residents, and 66% of rural residents—are of this opinion. Finally, central

TABLE 1.3   Attitudes Toward Community by Place of Residence
(percentages)

| How Do You Feel About This City/Neighborhood/ Community as a Place to Live? | City Residents | | Suburban Residents | Rural Residents |
	Attitude Toward Their City	Attitude Toward Their Neighborhood	Attitude Toward Their Community	Attitude Toward Their Community
Delighted/pleased or mostly satisfied	71	77	88	87
Mixed	17	13	8	8
Mostly dissatisfied or terrible/unhappy	12	10	4	5
Not sure	—	—	—	—
Total	100	100	100	100
Number of respondents	(3,298)	(3,298)	(3,229)	(547)

city residents are generally positive in their evaluations of the quality of
public services and facilities provided by local government. With the
notable exceptions of schools and road and street maintenance, majorities
believe that public services—including police and fire protection, garbage
collection, street lighting, parks and playgrounds, public transportation
and public health services—are either "excellent" or "good."

The City in Perspective

As indicated above, central city residents are relatively satisfied with
local public services. When such services are evaluated one by one, central
city residents are neither consistently more nor less satisfied than suburban
or nonmetropolitan residents with the public services available in each
type of place. City residents, for example, are more positive than their
noncity counterparts in regard to the quality of public health services,
street lighting, and public transportation; but they are more negative in
their evaluations of public schools and police protection—services that are
frequently important to residential satisfaction. Furthermore, when asked
to consider public services as a group and to assess the changes that have
taken place or will take place in the future, central city residents are
substantially more negative and pessimistic than persons living in other
types of places. People residing outside of the nation's larger cities, and
especially in small towns and rural areas, tend to be more positive about

the quality of public services. In some respects, then, the city is perceived in positive terms, but in other important respects (such as perceived quality of public services and schools), the city's image is much less positive in comparison to noncity locations (see Table 1.4).

A second example of offsetting images relates to the city's reputation as a place for employment and advancement. In addition to being seen as offering the best employment opportunities, cities have often been viewed as offering the most opportunities for advancement. The latter image, however, appears to be declining. Eighty-three percent of Americans believe that when they were growing up, the best advancement opportunities were found in the nation's cities; only 56% believe this is true of today's cities.

Finally, when one examines attitudes toward the physical and social environments provided by cities, the picture is also mixed. Certain aspects of urban life—such as the condition of streets, buildings, housing, and parks, the amount of crowding and congestion, or the leisure time facilities available—receive more positive than negative ratings from city residents. In comparison, noncity residents are generally more positive than are city residents about these same aspects of their communities. More importantly, attitudes toward these conditions are frequently less salient than attitudes toward other more obtrusive social conditions such as crime, drug addiction, and teenage gangs. On these scores, the city's reputation is poor and conditions are perceived to be getting worse with time. Percep-

TABLE 1.4   Ratings of Public Services by Place of Residence and Race (percentage saying services are "excellent" or "good")

| | | Place of Residence | | |
| | | | SMSA | |
Public Services	National	Central City	Outside of Central City	Small Town/ Rural Area
Police protection	67	62	72	66
Garbage collection	74	74	78	69
Street lighting	62	69	62	57
Fire protection	81	81	84	76
Public schools	62	47	67	73
Parks and playgrounds	51	52	52	48
Road and street maintenance	45	39	53	41
Public transportation	31	51	29	15
Public health services in hospitals and clinics	56	62	55	53
Number of respondents	(7,074)	(3,298)	(3,229)	(547)

tions of urban conditions and changing conditions are important because they may have an impact on migration patterns. It is to these patterns that we turn our attention.

MOVING TO AND FROM THE CITY

Not only have American central cities been losing population, there are indications from the HUD survey that this trend will continue for at least the next several years. The basic mobility patterns reported earlier must be examined further to determine who the movers are, why they are moving, and the impact of these movements on the composition of central cities. This section will describe the socioeconomic and demographic character-istics of those who have, in the recent past, and who may, in the near future, move to and from the nation's central cities.

Recent Movers to and from the City

We shall be examining data on the socioeconomic and demographic characteristics of each group of movers—those moving to the city from noncity locations and those moving away from cities. Since there is more movement out of the city than in the reverse direction, however, these data do not capture the net direction of change for each characteristic.

In general, cities appear to have been attracting a mix of people in recent years that is basically similar to their current population mix (see Table 1.5). Cities, more than other areas, have somewhat higher propor-tions of people who classify themselves as working class, of unemployed persons, of lower-income persons, of younger people, of single persons, and of blacks. Among recent movers to the city, there are higher propor-tions of these same types of persons than among recent movers out of the city. For example, 44% of all city residents classify themselves as working or lower class compared to 36% of all persons who live elsewhere. Among recent movers to the city, 47% classify themselves as working or lower class while 33% of movers from cities place themselves in these categories. Likewise, 22% of all city residents are single compared to 14% of noncen-tral city residents; 34% of recent movers to cities are single compared to 15% of recent movers from cities.

There are, however, some interesting variations. A much higher propor-tion of recent movers to cities are young and believe that their future mobility will be in an upward direction than either movers out of the city *or* of the city population as a whole. Although the data are not conclusive on this point, this portrait is consistent with the traditional image of the city as a vehicle for transformation and upward mobility of relatively

TABLE 1.5   Demographic Charactersitics of Recent Movers and Current City[a] and Noncity Residents (percentages)

	Recent Movers to City	Recent Movers from City	Current City Residents	Current Noncity Residents
Income Status				
Under $10,00C	41	28	39	29
$10,000 to $20,000	40	42	38	40
Over $20,000	19	30	23	31
Marital Status				
Single	34	15	22	14
Married	57	75	56	71
Widowed	2	3	12	10
Divorced	4	5	6	4
Separated	4	2	4	1
Social Class Status				
Upper/upper middle	23	34	29	37
Lower middle	30	33	27	27
Working/lower	47	33	44	36
Racial Status				
Black	14	5	24	6
Non-Black	86	95	76	94
Number of respondents	(151)	(296)	(2,132)	(4,834)

a. "City" refers to a central city of a Standard Metropolitan Statistical Area.

young refugees from depressed rural areas, whose children then move outward to the suburbs. As will be seen below, career mobility is one of the major reported attractions of cities. A more surprising variation is that although a larger proportion of recent movers into, as compared with out of the city, are black, the percentage of recent movers to the city who are black is smaller than the proportion of blacks who currently reside in central cities.

Recent Movers to Cities Who Want To Move Out

Another indication about the likely future size and composition of cities comes from data on people who have recently moved to cities and who now say that they may move out in the next several years. As many as 38% of recent movers to cities report that a move out is possible in the near future. In comparison, 10% of recent movers from large cities to noncity areas say that they may move back to the city in the next several years. For whatever reasons, movers to the city report that they are less likely than those who leave the city to remain permanent residents in their new location. Once again, this confirms a familiar demographic pattern—that moves are longitudinally linked and tend to be toward areas of perceived greater opportunity, amenities, and status.

Although the numbers of recent movers to and from the city in the HUD sample are relatively small, it appears that the recent movers to the city who are most likely to move out are white, young people and individuals with the least amount of education. Forty-four percent of the whites, 43% of the people under thirty, and 44% of those with less than a high school education who recently moved to the city indicate that a move for them out of the city is probable or definite. In contrast, only 10% of the whites, 13% of the people under thirty, and 8% of those with less than a high school education who recently moved from the city to a suburban or rural community indicate a preference to return to the city. Since many people moved to the city for jobs, it may be that either they were unsuccessful in fulfilling their career expectations or that they have been successful and now perceive "suburbia" to be a more attractive place of residence. It is important to place in perspective, therefore, the fact that some of the people who have been a part of the movement to the city indicate a preference to leave, while only a few of those who recently left the city intend to return. One of the implications is that urban revitalization efforts must do more than simply attract new residents to the city. They must also attempt more successfully to retain new arrivals.

Future Movers to and from the City

One way to get a rough notion of whether past trends will continue into the future is to ask people about their residential and mobility plans: do they intend to stay where they are or to move to another location? Of course, people's statements about the likelihood that they will move from their present residence may not result in actual mobility. However, if the projected time period is relatively short, say the next several years, then the assertion that a move is likely can give a reasonably good suggestion of the extent of future population shifts and the character of those shifts.

Respondents to the HUD survey were asked whether they "probably" or "definitely" would move from their present residence in the next two years and, if the answer was affirmative, where they would go. Thirty percent of all persons believed that some type of move was at least probable. An estimated 4.6% of the total survey population indicated that they would move from a central city to a nonlarge city area compared to 2.3% who indicated a move from noncentral cities to a "large city." How do the characteristics of these "future movers" compare with those who have recently moved to and from cities?

The socioeconomic profile of those noncentral city residents who say they are likely to move to the city in the next several years is somewhat different from profiles of current city residents, recent movers to cities, or future movers out of cities. The proportion of higher socioeconomic-status persons among potential future movers to the city is particularly noteworthy and surprising. While 29% of potential future movers out of cities classify themselves as upper-middle or upper class, 46% of potential movers to cities use these designations to describe themselves. Applying an objective criterion of social class, 32% of all potential movers to the city earn more than $20,000 per year in total family income compared to 23% of all city residents and 19% of recent movers to cities. And, compared to 15% of city residents and 22% of recent movers to cities who have at least a college degree, 30% of future movers to cities have attained this higher level of education. The group of future movers to the city also contains a higher percentage of younger and single persons than these other groups. On the other side of the coin, the group contemplating a move out of the city in the near future contains a higher proportion of unemployed, lower-income, and less educated persons than the group of potential movers to the city.

THE CHANGING CITY

Information about the proportion of future movers to the city who have higher socioeconomic characteristics has led some observers to conclude that cities are experiencing or are about to experience a turnabout in population trends. However, this evidence must be examined in context. Since more people indicate a future intention to move out of rather than to cities, it is necessary to consider the net gain or loss from migration that is likely to occur for different socioeconomic and demographic groupings. This section examines the size and composition of cities that would result from the moves that are being contemplated during the next several years.

The first outcome seems obvious: cities will be smaller than they are at present, with a continuation of the recent trend of net out-migration. Approximately 15% of all persons who presently reside in central cities intend to move in the next few years to areas outside of such cities. In contrast, only 3% of all persons who presently reside outside of central cities plan to move to a large city within the next few years. These percentages, of course, must be evaluated from the perspective of the relative sizes of the respective populations. Given that there is an approximate 7:3 ratio between noncentral city and central city residents, the net directional flow is one of continued population decentralization into the suburbs and outlying areas. By taking into account the number of persons intending to move out of and into the nation's central cities, it can be projected that, on balance, cities will lose approximately 9% of their total population or almost 5 million people in the next few years through migration. Hence, an analysis of moving intentions provides no evidence to indicate that the cycle of population decline that has characterized cities since the 1960s will reverse itself in the near future.

This analysis can also focus on the type of people who will reside in cities in the near future. On the basis of moving intentions, it can be projected that the basic socioeconomic and demographic characteristics of cities will *not* change dramatically during this period of continued population decline. However, it does appear as if cities, in the next few years, will have a slightly larger proportion of professionals, people over thirty, retired individuals, and people who consider themselves to be upper-middle class. The projected slight increase in the percentage of professionals who will be residing in the city in the near future is probably the most encouraging sign for the economic well-being of cities. Nevertheless, even this positive development must be evaluated in context, since the total *number* of professionals residing in cities will be declining while the relative *proportion* of professionals in cities will be increasing. It can also be projected, based on moving intentions, that the approximate 3:1 ratio between whites and minorities in central cities will not change dramatically in the process of population decline. The short-term result of moves into and out of central cities, then, will be a net decline in population but a demographic and racial mix that is approximately the same as the mix that exists today.

THE ATTITUDES AND CONCERNS OF MOVERS

In considering various urban revitalization strategies, it is useful to ask what attitudes are associated with moving and what reasons relate to a move. This section examines some of the attitudes of those who say they are likely to move either into or out of cities and compares such persons to nonmovers in their present type of community.

PERCEIVED COMMUNITY PROBLEMS AND DISSATISFACTIONS

City residents who anticipate leaving for a nonlarge city location in the next several years share with the majority of city residents the belief that crime, unemployment, and drug addiction are particularly severe in the city. These views, therefore, do not distinguish them from those who probably will stay in the city. There are some attitudes, however, which do

distinguish movers from nonmovers and these fall into three broad categories: perceptions about various adverse physical and environmental conditions, concerns about personal safety, and attitudes related to child
rearing. Potential movers *from* central cities are somewhat more likely
than nonmovers to see conditions like traffic congestion, air pollution, and
noise levels in the city as severe problems. In the case of air pollution,
about half of all movers compared to 36% of nonmovers believe it to be a
problem. Movers and nonmovers differ even more in their concern about
personal safety and child-rearing conditions of the city. Compared to 28%
of all nonmovers who are dissatisfied with the safety afforded them in the
city, 45% of all potential movers away from cities register a similar amount
of dissatisfaction. Twice as many movers as nonmovers are dissatisfied
with the city as a place to raise children. Compared to 40% of all
nonmovers who rate the city's public schools as "fair" or "poor," 56% of
all prospective movers register a similar attitude.

In contrast, potential movers to the city from suburban or nonmetropolitan areas have a different set of attitudes toward their current locations. Compared to those who intend to stay in a noncity location,
potential movers to the city are more frequently dissatisfied with the
recreational, entertainment, and leisure-time activities available in their
communities, with the beauty and attractiveness of their immediate environment, and with a general lack of interesting things to do. For
example, 49% of those who intend to move to the city identify the lack of
interesting things to do as a "severe" problem compared to 23% of those
who anticipate staying where they currently live.

Potential movers from suburban and nonmetropolitan areas share with
most Americans some of the more negative attitudes about the nation's
cities: 93% of all adult Americans and 97% of potential movers to cities
believe that cities have more crime than any other type of place. Those
who may move to the city, however, exhibit a greater sense of optimism
about other attributes of city life than do noncity residents who contemplate no such move. For example, 18% of noncity nonmovers say that
cities have the best public schools; 40% of potential movers to the city
express this belief. Only 7% of those who intend to remain out of the city
see cities as the best places to raise children; 21% of potential movers to
the city assert its child-raising advantages. Compared to 8% of noncity
nonmovers, 31% of potential movers to the city believe that cities contain
friendlier people than other types of places.

WHY PEOPLE MOVE

People move for many reasons and it is important to understand why
people are attracted to or repelled from the contemporary city. The

reasons most frequently given for moving *to* cities relate to employment opportunities and amenities; the reasons given for moving *from* cities relate to a desire to improve one's quality of life through the upgrading of neighborhoods and housing.

A majority of people indicating a preference to move from suburban or nonmetropolitan areas to central cities cited job or career-related concerns as their primary reason. Sixty-one percent of the people who anticipated moving to a city mentioned an employment-related reason, followed by 37% who cited community amenities. These figures indicate that the city's image as a center for jobs and career advancement and amenities is still the major attraction to new residents (see Table 1.6).

Single persons (80%), those who completed at least high school (70%), and people under thirty (70%) are much more likely to cite employment as a reason to move to the city than married people (41%), those with less than a high school education (29%), and people over thirty (51%). People with higher education are more likely to be attracted to the job or career opportunities perceived to be available in cities than those with less education. Persons with less than a high school degree mention housing quality considerations (36%), neighborhood conditions (32%), and jobs (29%) when asked why they wanted to move to the city, while, in contrast, those who have completed high school (or beyond) cite jobs (70%), community amenities (41%), and neighborhood conditions (18%) as motivating factors. These findings may suggest a change in the types of jobs cities are perceived to offer; they may now be seen as having fewer of

TABLE 1.6   Reasons for Wanting to Move to or from Central Cities[a]

Reasons	People Intending to Move to Cities	People Intending to Move from Cities
Job or career change/opportunities	61	26
Housing quality or condition	14	30
Friends/relatives/neighbors	6	4
Amenities/environment	37	36
Schools	2	12
Moving away from parents/getting married	3	2
Retiring	2	4
Cost of living/prices	7	11
Neighborhood conditions	14	40
Number of respondents	(153)	(322)

a. Figures in the table are percentages of respondents who cited each specific item as one of the most important reasons why they intended to move. Each respondent was permitted to give a maximum of three responses to the question.

the traditional lower-wage, entry-level jobs that attracted low-status migrants in past decades.

Those who want to move *from* the city to nonurban areas are much less likely to cite jobs or careers as reasons than people who intend to move *to* the city. City residents most frequently cited negative aspects of their existing neighborhoods or a desire to move to better neighborhoods as reasons why they might be leaving the city. Specifically, neighborhood conditions (40%), community amenities (36%), housing quality (30%), and careers (26%) are the most frequently stated reasons for wanting to leave. Although "neighborhood" considerations dominate, these may, in fact, incorporate many different types of perceived conditions including physical deterioration, declining property values in the surrounding area, social or racial tensions, inferior educational facilities in the vicinity, and so forth.

Reasons for wanting to leave the city do not appear to be associated with marital status or education. However, there are some differences across age groups. People under thirty, for example, mentioned community amenities (34%), jobs (33%), and housing quality (30%) as the reasons why they intended to leave the city. But people over thirty cited neighborhood conditions (41%), community amenities (37%) and housing quality aspects (31%) as their major reasons for desiring to move out of the city.

IMPLICATIONS

The attitudinal and demographic patterns discussed above have implications for the strategies that are adopted to revitalize the nation's urban areas and for the possibility of achieving urban revitalization. Revitalization programs either explicitly or implicitly assume the existence of certain public preferences or the likelihood of certain public behaviors in response to actions taken by government, community-based organizations, business groups, etc. The contemporary debate on the subject of urban revitalization, however, suggests that there is disagreement about these preferences and behaviors. For example, there is disagreement as to:

(1) whether the postwar trend of net out-migration from central cities has begun to stabilize or to reverse itself;

(2) whether "urban" problems have or are perceived to have spread to the suburbs, thereby diminishing the perceived advantages of suburbs over cities as desirable residential environments;

(3) whether the middle class is "returning" to the cities and, if so, whether in sufficient numbers to offset the movement of middle-class persons out of cities; and

(4) whether improving physical or social conditions, upgrading existing amenities, facilities or services, or creating new career opportunities are the most appropriate means of retaining existing or attracting new residents to the cities.

This chapter has begun to address questions such as these with data from a comprehensive public survey on the quality of community life. Several points have been made:

(1) There is little evidence on a national scale of a net shift in population into cities, and there is little evidence that such a shift can be expected in the near future. Based on people's reported moving intentions, central cities will continue to lose population through migration, but the demographic and racial characteristics of cities will not change dramatically.

(2) Problems related to the social and residential environments—crime, drug addiction, teenage gangs, poor housing, dirty streets, inferior schools, and high unemployment—are still perceived by the public to be severe problems associated with cities and not with suburban or rural areas. Although it may be true that some of these problems may be overflowing into suburban communities, the public continues to associate these types of problems primarily with large cities. Moreover, there is no evidence that large numbers of Americans now see cities as the preferable residential location because of the spreading of social problems to noncities.

(3) A greater proportion of potential movers to the city are higher income and better educated than those who have recently moved to the city or who are currently residents of the city. This development, which may be linked to the changing occupational structure of cities, could signal a reversal in migration patterns and, if so, it would have far-reaching effects. On the other hand, although the *proportion* of upper-socioeconomic-status persons residing in cities may increase slightly in the next few years, the total *number* of such people living in cities will be declining at the same time.

(4) People intending to move to the city tend primarily to be attracted to its perceived employment opportunities and amenities, while those who may be leaving the city are either repelled by negative neighborhood conditions or are attracted by the positive, nonurban neighborhood characteristics, amenities, or housing conditions of other places of residence.

The implications of these findings depend on the goals of urban revitalization. If the revitalization focus is on restoring the city's important role as an engine for raising up and transforming economically deprived and new immigrant populations, then the focus should be on increasing jobs and training appropriate to these groups' needs. Alternatively, if the revitalization goal is to attract more middle-class residents to the city, then the focus should be on reducing the disparity in quality of life (schools, crime, etc.) between suburbs and cities.

The challenge of the latter focus involves not only changing the actual conditions of cities but also altering the negative imagery associated with them. The attitudinal data presented in this paper suggest that there are three distinct ways of looking at this imagery. First, there are certain overtly negative perceptions of cities about which there is a fair amount of public consensus. Second, certain judgments about cities that may appear to be positive should be seen in the context of similar perceptions of suburban, small town and rural communities; these comparisons often show the city as less attractive or satisfying than noncity alternatives. Third, the city should be viewed in the context of its changing reputation; again, certain attributes that may appear to be judged as positive, by themselves, seem less so when examined against the public's sense of how conditions have been changing over time.

Considered in these contexts, the nation's large cities are clearly stigmatized. From a revitalization perspective, this is significant because negative images can take on a reality of their own and, thus, help to perpetuate many of the social, economic, and fiscal problems that cities face.

REFERENCES

ALLMAN, T.D. (1978) "The urban crisis leaves town." Harper's Magazine 256 (December): 41-56.

BERRY, B. and D. DAHMANN (1977) Population Redistribution in the United States in the 1970s. Washington, DC: National Academy of Sciences.

Congressional Quarterly (1979) "Back to the cities?" 32 (January 6): 26.

FLEETWOOD, B. (1979) "The new urban elite and urban renaissance." New York Times Magazine (January 14).

NATHAN, R. and J. FOSSETT (1978) "Urban conditions—the future of the federal role." Presented at meeting of National Tax Association, Philadelphia, November.

NELSON, R. (1979) "Recent suburbanization of Blacks: how much, who, and where." Washington, DC: U.S. Department of Housing and Urban Development.

PETERSON, G. (1976) "Finance," pp. 35-118 in W. Gorham and N. Glazer (eds.) The Urban Predicament. Washington, DC: Urban Institute.

U.S. Congress, Joint Economic Committee (1977) The Current Fiscal Condition of Cities: A Survey of 67 of the 75 Largest Cities. Washington, DC: Author.

U.S. Department of Housing and Urban Development (1978a) Office of Policy Development and Research. A Survey of Citizen Views and Concerns About Urban Life. Washington, DC: Author (May).

——— (1978b) The President's National Urban Policy Report. Washington, DC: Author (June).

——— (1978c) Office of Policy Development and Research. The 1978 HUD Survey of the Quality of Community Life: A Data Book. Washington, DC: Author (November).

——— (1978d) Office of Community Planning and Development. Whither or Whether Urban Distress. Washington, DC: Author (December).

——— (1979a) Office of Policy Development and Research. Urban Fiscal Crisis: Fantasy or Fact. Washington, DC: Author (March).

——— (1979b) Office of Policy Development and Research. Changing Conditions in Large Metropolitan Areas. Urban Data Reports, No. 1. Washington, DC: Author (June).

YENTIS, D. (1978) 'The 'back to the city' movement—fact or fiction." Unpublished Manuscript.

NOTE

1. Estimated maximum sampling error for the total sample, including the effect of clustering is 2.4 percentage points. For the city, suburban and nonmetropolitan strata, it is 2 percentage points, 2 percentage points, and 4.9 percentage points, respectively. For additional information regarding the survey, see U.S. Department of Housing and Urban Development, 1978a, 1978c.

2

The Costs and Benefits of
Neighborhood Revitalization

BILIANA CICIN-SAIN

INTRODUCTION

□ BOTH POLICY-MAKERS AND URBAN ANALYSTS have for some
time recognized that one of the major reasons for the decline of America's
cities during the postwar period was the loss of the urban middle class.
Encouraged by a variety of local, state and federal government policies,
large numbers of middle- and upper-income families chose to live in
suburban communities rather than in central cities. The result of these
movements has been a series of socioeconomic disparities between central
cities and the remainder of metropolitan areas. As characterized a few
years back, "One set of jurisdictions—central cities—has the problems, the
other—the suburbs—has the resources" (Lineberry and Sharkansky, 1974:
28).

For the first time, we appear to be witnessing a reversal of these trends
which suggests improved prospects for central city neighborhoods and
perhaps also marks the end of a long period of urban decline. Variously

AUTHOR'S NOTE: *This chapter is based on a paper entitled "Displacement as a
Result of Neighborhood Revitalization: The Policy Issues and the Research Issues"
prepared by the author while she was a National Association of Schools of Public
Affairs and Administration (NASPAA) Faculty Fellow in the Office of Policy
Development and Research in the U.S. Department of Housing and Urban Develop-
ment (HUD) in January 1978. A version of that paper (revised by Howard J. Sumka,
a HUD official) has been published as "Displacement in Revitalizing Neighborhoods:
A Review and Research Strategy," in* Occasional Papers in Housing and Community
Affairs, U.S. Dept. of Housing and Urban Development, Vol. 2, 1979. *(See errata
listed in Volume 4, 1979.)*

labeled "gentrification," "the back-to-the-city movement," "return of the chic," "urban renaissance," the process of neighborhood revitalization has begun to capture the attention of the nation's media (Hartnett, 1977; Kornegay, 1977; Reinhold, 1977; Ross, 1977; Boston Globe, 1977). Increasingly, there is evidence of neighborhood revitalization activity in central city neighborhoods across the country resulting largely from a variety of private market forces and demographic and life-style changes, and, to some extent, also to government policies. While it is clear that some activity is underway in certain neighborhoods, and there are indications that such activity is likely to expand in the future, nevertheless there remain important unanswered questions about the exact extent and nature of the national neighborhood revitalization effort.

Increased renovation seems to be primarily due to two major forces. First, the growing shortage of affordable housing in many metropolitan areas (due to escalating housing costs, federal building moratoria of the early 1970s, and the growth controls and exclusionary actions of suburban governments) has induced many to consider lower-cost inner city housing alternatives. Second, this housing market decline has coincided with the entry into the housing market of large numbers from the postwar baby boom generation. This numerically dominant generation has pioneered new life-style trends and new patterns of household formation which are often more amenable to central city residence than to traditional suburban living patterns (James, 1977a). Related factors which may contribute to the spur in renovation include the energy crisis, which has considerably dampened the lure of commuting long distances to suburban locations, and the concerted efforts of city officials to halt a declining tax base.

A basic public policy question that is posed by this revitalization phenomenon is: What are the costs and benefits of neighborhood revitalization for different categories of individuals, social and ethnic groups, neighborhoods, cities, metropolitan areas, and, ultimately, for the nation as a whole? Conceivably, the process of neighborhood revitalization could result in a variety of different outcomes or scenarios. On the one hand, neighborhood revitalization could yield healthy, diverse communities with a mix of race, age, and income. It could help cities to achieve many long-standing goals, such as improving the housing stock, increasing the tax base, keeping or attracting middle- and upper-income households to the city, bringing back businesses, and improving the quality of services delivered. At the same time, revitalization could be accompanied by significant social costs. Revitalization of central city neighborhoods by upper-social-status newcomers could merely work to shift intractable problems of poverty, unemployment, and inadequate housing to other parts of a metropolitan area through the wholesale displacement of the old, the poor, and minority residents.

Most of the available information on neighborhood revitalization is based on single case studies of particular neighborhoods in particular cities. Such data are largely unsystematic and impressionistic and often run the risk of being colored by the a priori biases of the observer. The result has been a somewhat unbalanced treatment of the subject—revitalization alternately being praised as the salvation of urban America or being condemned as a new white middle-class plot to drive out the black, the poor, and the Hispanic.

The purpose of this chapter is threefold. It first reviews the scant empirical evidence available on the process of neighborhood revitalization, focusing on its incidence, characteristics, and human dimensions. Second, it sets forth an analytical framework for considering the potential or hypothetical consequences of revitalization on affected parties. While costs and benefits are identified only in conceptual and hypothetical terms (given the absence of data), this exercise is useful insofar as it reveals how the mix of costs and benefits of revitalization is likely to vary according to whose perspective is being considered. Third, the chapter considers the evidence available on the most negative aspect of revitalization—the displacement of lower-income residents from their neighborhoods.

THE PROCESS OF NEIGHBORHOOD REVITALIZATION

The incidence of privately induced neighborhood revitalization is a matter of considerable controversy. As Abravanel and Mancini report in this volume and others elsewhere (U.S. Department of Housing and Urban Development 1979a), national-level data continue to suggest that population migration patterns are still sapping the economic and social vitality of cities. This is particularly true of the older urban cores of the Northeast, victims of the recent Sunbelt phenomenon (Sternlieb and Hughes, 1977). At the same time, a growing body of literature—albeit much of it anecdotal—suggests that significant neighborhood revitalization activity is taking place.

A number of multicity studies document the extent of neighborhood renovation. A mail survey administered in 1975 by the Urban Land Institute (ULI) to housing and planning officials in every central city in the nation (Black, 1975) found that 48% of central cities with a population of over 50,000 were experiencing some degree of private-market, nonsubsidized housing renovation in older, deteriorated areas. The incidence of renovation was found to vary considerably according to size, with larger cities experiencing this type of renovation activity to a greater extent (73% of cities in the 500,000 and over population group were experiencing renovation versus 32% in the 50,000 to 100,000 class).

Quantitatively, the ULI survey estimated that perhaps as many as 50,000 units had been renovated in these revitalizing neighborhoods between 1968 and 1975. A 1976 mail survey of 43 cities conducted by the National Urban Coalition found that

> the phenomenon of housing rehabilitation is widespread, regardless of city size or geographic location. The scale of this movement varies. In some of the larger cities . . . surveyed—San Francisco, New Orleans, Washington, D.C., Denver—the movement might be described as extensive and aggressive—where seven to ten neighborhoods at one time are at various stages in the revitalization process. But even in the smaller jurisdictions, Newburgh, New York, for example, or Bridgeport, Connecticut, a few determined residents are spurring private reinvestment [Holman, 1977: 189].[1]

Additional evidence of heightened renovation activity is provided by a 1977 Urban Institute study by Franklin J. James who reviews national population and housing trends in metropolitan areas across the country for the period 1970-1975 using such data as the Annual Housing Survey and the Survey of Residential Repairs and Alterations. His analysis reveals several manifestations of a national renovation trend, i.e., for the first time, in 1973-1975, median housing values and median gross rents increased faster in central cities than in suburbs; small but significant increases in homeownership patterns were observed in central cities during the 1970-1975 period; and, during the same period, median home improvement expenditures for central city homeowners rose abruptly and for the first time in a long time exceeded suburban expenditure rates (James, 1977a: Ch. 2). Another 1977 study of the largest U.S. cities conducted by S. Gregory Lipton at the University of Oregon found that in a number of cities, the proportion of middle- and upper-income families living in census tracts within two miles of the central business district had increased between 1960 and 1970 (Lipton, 1977).

In addition to these multicity studies, there exists a growing volume of anecdotal evidence on the renovation process in different neighborhoods in central cities across the country. While most of these case studies are based on impressionistic accounts of the renovation process by media observers, public officials, academics, neighborhood groups, and neighborhood residents (and hence the data are largely noncomparable), the totality of the evidence indicates that a significant amount of heightened renovation and residential upgrading activity is taking place. There are indications, too, that much of this activity is relatively recent; most of it is reported to have taken place only within the last three to five years. As Carl Holman of the Urban Coalition puts it, "Where once it took five years or more to 'turn around' a neighborhood, our survey reveals that in neighborhoods under significant renovation pressure, this may be accomplished in 24 months or less" (Holman, 1977: 190).

While, the neighborhood revitalization phenomenon is characterized by a great deal of diversity, the available data suggest some commonalities. Revitalization seems more likely to occur in: (1) larger cities (over 100,000) (Black, 1975); (2) in the Northeast and South (Black, 1975); (3) in centrally located neighborhoods which are close to the central business district or are near convenient mass transit facilities (Schur, 1977; Lipton, 1977; Holman, 1977); (4) in neighborhoods which have received historic designation or which are close to historically designated areas (Holman, 1977); (5) in neighborhoods in which there is good housing stock available, that is, housing stock which no matter how old or dilapidated was originally built to luxury or at least to upper-income standards for its time (Schur, 1977); (6) in neighborhoods which are close to an area of physical beauty or other focal points of interest, such as a university (Holman, 1977); and (7) in neighborhoods in which infrastructure amenities such as parks, good schools, libraries, and hospitals exist (Schur, 1977).

Neighborhood revitalization also seems to vary according to which actors are spearheading the activity. Two major categories of cases are evident: renovation which is undertaken by present residents either with assistance of public funds (as in Boston and Mount Auburn in Cincinnati) or without such assistance (as in Pittsburgh, Detroit, Minneapolis, and Milwaukee); and renovation which is undertaken by generally higher socioeconomic status newcomers—as has been the case in Adams-Morgan, Mount Pleasant, and Capitol Hill in Washington, DC (Pattison, 1977).

Who are the newcomer residents who are revitalizing these neighborhoods across the country? While the most dramatic examples of population change in neighborhoods undergoing revitalization have involved a transition of black to white residents, it appears that the crucial factor underlying population shifts in neighborhood renovation cases has been socioeconomic status. As the Urban Coalition study reveals:

> In almost every neighborhood surveyed, with the exception of Neighborhood Housing Service Areas, the crucial difference between residents before and after renovation was their social and economic background. Race, age, family size, type of tenancy, may and frequently do correlate with socio-economic class, but these factors are secondary [Holman, 1977: 191].

This notion is further reinforced when one takes note of several examples of renovation in black neighborhoods by incumbent residents— e.g., Mattapan in Boston, University City in St. Louis (James, 1977b), and when one notes that in the period 1970 to 1974 black homeowners increased their central city home repair activity (both in terms of projects and expenditures) more than did white central city homeowners (Embry, 1977).

What are the characteristics of these higher-status newcomers? The national media has painted a composite profile of housing renovators as being primarily young (predominantly in the 25-to-44 age category), as mainly engaged in professional occupations, predominantly affluent, largely childless, and mostly falling into the working couple category or into single and divorced statuses. The media has also tended to cast these "young professionals" as part of a "back-to-the-city movement" which rejects the monotony, dullness, and energy waste of suburbia in favor of the central place attractiveness and energy efficiency which characterize central cities.

By and large, the available data support the stereotype presented by the media (Black, 1975; Gale, 1976, 1977; Pattison 1977; James, 1977b) with one significant exception: instead of a "back-to-the-city movement" we may be witnessing a "stay-in-the-city" movement. For the available evidence suggests that the newcomer residents in neighborhoods undergoing revitalization (at least among owner-occupants) have, by and large, moved from within the central city. In the Urban Institute study, for example, James (1977a) showed that 70% of households that purchased homes in central cities in 1973 and 1974 were locating within the same central city. Of these, two-thirds were renters moving into homeownership for the first time. An additional 13% of homeowners were relocating from other central cities or from nonmetropolitan areas. Very few buyers were moving from the suburbs. Only 11% of central city homeowners were moving from the suburbs to the city. Thus, James concludes that "Clearly, there is no evidence of a back-to-the-city movement. It is much more apposite to view the revitalization of demand for homes in cities as the result of the changing housing needs and changing housing constraints on city residents" (James, 1977a: 165).

Other studies support this general conclusion. Rogg (1977), for example, found that of 140 families moving into the Baltimore Neighborhood Services rehabilitation project area, 75% were residents of the area and only 25% were from outside the area. Gale (1977: 3) in his study of Mount Pleasant, also found that

> The large majority of new Mount Pleasant homeowners had lived in the District of Columbia just prior to moving to their new neighborhood. Less than one-fifth, however, had moved to the neighborhood from D.C. suburbs. Thus, for the most part, the renovation activity there does not appear to represent a significant gain in population to the District. . . . About half had lived in apartments and more than two-thirds were renters before buying homes in Mount Pleasant.

BILIANA CICIN-SAIN   55

Similarly, in his study of West Cambridge, Pattison (1977) found that the overwhelming majority of new owner-occupants had moved from within the city. The available data (although spotty) thus suggest that at least among the owner-occupants, neighborhood renovation might more accurately be described as a "stay-in-the-city" phenomenon rather than a "back-to-the-city movement."

Among renters, however, a different phenomenon may be operative. In a study of mover households into the District of Columbia, George and Eunice Grier (1977) showed that five out of six of these recent newcomers came to the District from completely outside the metropolitan area; only a small minority came from the suburbs. The overwhelming majority of newcomers chose rental units—primarily in close-in central locations undergoing private rehabilitation in the renewal areas, or in high-prestige areas on the west side of the city.

Given the predominantly young age of newcomers in neighborhoods undergoing revitalization, many argue that neighborhood revitalization might be a temporary phenomenon. As the newcomers age and begin to have children (or have more children) it is argued that long-standing suburban attractions such as good schools and open space will draw them away from central city locations. Although very few studies address this question, the limited case study data that are available (Gale, 1977; Pattison, 1977) suggest that this is not the case and that generally, the newcomers tend to be committed urbanites.

NEIGHBORHOOD REVITALIZATION: THEORETICAL COSTS AND BENEFITS

A useful means of organizing our thinking about the potential effects of neighborhood revitalization is a cost-benefit matrix which explicitly identifies the affected parties and revitalization's effects on them. Given the absence of empirical data on the *actual* effects of revitalization at this point, one must identify costs and benefits solely in *conceptual* and *hypothetical* terms. Nevertheless, the matrix is useful as it focuses attention on problematic areas (e.g., what groups are likely to be most detrimentally affected) and pinpoints specific research questions and specific policy issues that will need to be addressed.

The cost-benefit framework (summarized in Table 2.1) identifies several relevant units of analysis, i.e., individuals, groups, neighborhoods, cities, metropolitan areas, and the nation as a whole. An effort has been made to identify all of the potential effects which may accrue to each of these units, without regard to the practical problems associated with empirical measurement. Potential effects are classified according to whether they are *likely* to be positive ("potential benefits"), or negative ("potential costs"), or uncertain—either positive or negative ("indeterminate effects"). The large number of entries which the reader will find categorized as "indeterminate effects" will underscore how little we actually know about revitalization and how uncertain we are, in the absence of empirical data, about whether particular effects will prove beneficial or detrimental, and for whom.

Table 2.1 first considers the mix of costs and benefits which may accrue at the *individual* level. Four major classes of individuals are identified here: individual residents who remain in the neighborhood following revitalization, those who leave, the newcomer residents who generate or sustain the renovation activity, and individuals involved in the real estate sector and in the building trades.

Those involved in real estate and building activity are most likely to be clear winners in the revitalization game as they spearhead much of the renovation activity and hence profit by the rising value of the housing stock, the increased number of transactions, and by the increased building business generated.

Remaining neighborhood residents face a mixed bag of costs and benefits. On the positive side, they stand to gain from an improvement in neighborhood conditions—in terms of housing stock, physical environment, increased municipal services, and upgrading in the quality of locally sold goods and services. Remaining homeowners can realize substantial equity increases on their property and stand to gain from the increased availability of financial services. On the other hand, these households are likely to meet increased housing costs (increased rents for those renting and increased taxes for homeowners), higher prices of locally sold goods and services and the loss of previous social and institutional ties. Rising costs as well as pressure to move resulting from harrassment by real estate brokers and landlords may ultimately force some to move out.

Former neighborhood residents—those who leave—comprise the category likely to be most adversely affected by revitalization. These people undergo considerable moving costs (both psychic and material) and the loss of established institutional and social networks. As we shall see below, little is known about their post-move circumstances—where they move to and how their new housing, neighborhood and living conditions compare to those they left behind. The only readily apparent benefit which accrues to this group is the recapture of equity appreciation for the former owners (provided, of course, that they did not sell their properties during the early stages of revitalization).

Among those leaving, individuals falling into particular categories (i.e., the poor, the aged, female-headed households and members of certain ethnic, and racial groups) may be negatively affected by revitalization in

TABLE 2.1 Potential Effects of Neighborhood Revitalization

Unit of Analysis	Potential Costs (−)	Indeterminate Effects (?)	Potential Benefits (+)
I. Individuals			
(A) Real Estate Sector and Building Industry			
(1) Investor			Increased profits
(2) Broker			Increased profits
(3) Building suppliers			Increased profits
(4) Builders			Increased profits
(B) Residents			
(1) Remaining Neighborhood Residents	Increased housing costs: —Owners: increased taxes —Renters: increased rent	Change in characteristics of neighborhood population	Equity appreciation for homeowners
	Subject to pressures to move through harassment by real estate brokers or landlords	Loss of old social ties/gain of new social ties	Improved municipal services
			Improved physical environment
	Increased costs for locally sold goods and services		Improved quality of local goods and services
			For owners: —improved availabillity of mortgage and home improvement credit —increased availability of hazard insurance

TABLE 2.1 Potential Effects of Neighborhood Revitalization (Cont)

Unit of Analysis	Potential Costs (−)	Indeterminate Effects (?)	Potential Benefits (+)
(2) Former Neighborhood Residents			
(a) in general	Moving costs (e.g., transportation costs, security deposits, new utilities, psychic costs). Also, *how* many times do they have to move?	Change in characteristics of new housing (size, cost, quality)	Homeowner recapture of equity appreciation
	Loss of old social and institutional ties	Change in nature of new neighborhood (in quality of municipal services, schools, availability and costs of transportation, proximity to employment (including availability of semiskilled and unskilled jobs in new neighborhood)	
	Trauma of forced move, particularly for renters		
		Change in accessibility to city amenities (arts, sports, recreation, culture)	
		Receptivity to newcomer by new neighborhood	
		Opportunity for integration in a suburban neighborhood	

TABLE 2.1 Potential Effects of Neighborhood Revitalization (Cont)

Unit of Analysis	Potential Costs (−)	Indeterminate Effects (?)	Potential Benefits (+)
(b) special categories[a]			
(1) the poor		Availability of unskilled and semiskilled employment opportunities in recipient communities**	
		May be detrimentally affected by limited growth policies of suburban governments (i.e., unable to settle in certain communities)*	
		Availability of special services for the poor (such as welfare offices, neighborhood centers, training centers, etc.)*	
(2) the aged	Availability of special services such as senior citizen services*	Elderly may also be disproportionately affected by loss of social ties**	
	Proximity to medical services (and transportation costs)*		

TABLE 2.1 Potential Effects of Neighborhood Revitalization (Cont)

Unit of Analysis	Potential Costs (−)	Indeterminate Effects (?)	Potential Benefits (+)
	May also be disproportionately affected economically because of fixed incomes**	May be unable to find suitable living quarters (in size)**	
(3) female headed households/large families	Female-headed households may experience discrimination in renting or buying*		
(4) ethnic and racial groups	Loss of political representation which some groups have been able to achieve as a result of numerical concentration*		
	May be particularly affected by loss of social and community ties based on ethnicity or race**		
	May suffer discrimination on the part of recipient communities*		

TABLE 2.1 Potential Effects of Neighborhood Revitalization (Cont)

Unit of Analysis	Potential Costs (−)	Indeterminate Effects (?)	Potential Benefits (+)
(3) New Neighborhood Residents	Adjustment to central city living	Change in social and institutional ties	Lower housing costs
			Homeowner equity accumulation
	Potential physical danger due to conflict with remaining residents	Adequacy and cost of municipal services (particularly schools for those with children)	Proximity to employment
	For owners: risk of equity loss if neighborhood does not stabilize		Sense of accomplishment/urban pioneer motive
			Proximity to city amenities (arts, culture, recreation, sports)
			Lower transportation costs
II. Neighborhoods			
(A) Revitalizing Neighborhood	Disruption of functioning neighborhood networks and institutions of self-help, social control, and planning	Loss of diversity in characteristics of the population	Upgrading of housing stock
	Creation of racial and social class tension (during revitalizing stage)		Purchases of rehab services and materials, creation of rehab jobs (it is uncertain though, whether these resources would accrue to the revitalizing neighborhood)

61

TABLE 2.1 Potential Effects of Neighborhood Revitalization (Cont)

Unit of Analysis	Potential Costs (−)	Indeterminate Effects (?)	Potential Benefits (+)
			Increase in quality of muncipal services
			Availability of private financing (i.e., redlining no longer a problem)
(B) Recipient Neighborhoods[b]	Creation of racial and social class tension	Availability of special services for special groups, e.g. aged, poor	
	Increased load on municipal services	Increased/decreased social homogeneity/diversity	
	Reduction in tax base (property, sales, income)		
III. Cities	Cost of improved services demanded by new residents	Decline in school enrollments (influenced by a decline in the number of schoolchildren)	Increased tax base (property, sales, income)
			Increased employment: – Real estate and building sectors – Other service sectors
			Improved muncipal services

TABLE 2.1 Potential Effects of Neighborhood Revitalization (Cont)

Unit of Analysis	Potential Costs (−)	Indeterminate Effects (?)	Potential Benefits (+)
			Decrease in cost of services required by low-income population
IV. Metropolitan Area as a Whole	Potential shift of the problems of poverty from central cities to other parts of the metropolitan area		Revitalized central city
			Redress of socioeconomic disparities between central city and suburbs
	Limited population growth policies by suburban communities may prevent displaces from entering		
	Increased service load on recipient communities		
	Potential resegregation and racial conflict		
V The Nation	Subsidy for residents wishing to remain in area	Uncertain impact on problems of poverty	Revitalized central cities
	Relocation assistance	Uncertain impact on metropolitan racial and economic resegregation	Conservation of existing housing stock and capital infrastructure

TABLE 2.1 Potential Effects of Neighborhood Revitalization (Cont)

Unit of Analysis	Potential Costs (−)	Indeterminate Effects (?)	Potential Benefits (+)
			Conservation of energy and land
			Restoration of local fiscal balance

a. Among those leaving, individuals falling into special categories (i.e., the poor, the aged, female headed households/large families and members of certain ethnic and racial groups) may be affected by revitalization in *additional* and/or *disproportionate* ways; "additional" (*) ways in the sense that: (1) they may lose particular advantages which they may have had by virtue of their concentrated numbers, or (2) they may suffer additional problems because of their status as members of a particular category; "disproportionate" (**) in the sense that they may bear certain costs (common to all those who are leaving) disproportionately by virtue of their special needs or lack of options. These are marked accordingly (* or **).
b. Neighborhoods into which displaced individuals are moving. Recipient neighborhoods may include: other city neighborhoods, suburban municipalities, unincorporated areas. The potential effects listed may not accrue to all types of recipient communities.

additional and/or disproportionate ways. The lower the income of those leaving, the greater will be the negative impact of dislocation on them. In large part, this results from normal market operations which leave the poor with fewer choices than wealthier families. This effect may be compounded by the limited growth policies and exclusionary practices of suburban communities. Low-income families will be particularly affected if a forced move diminishes their accessibility to opportunities for unskilled and semiskilled employment. Equally distressing for them may be the unavailability of special services which tend to be supplied in and around traditionally low-income areas, such as welfare offices, neighborhood service centers, and job-training centers.

Certain demographic groups (female-headed, large and elderly households) may find dislocation particularly traumatic. Female-headed households may experience discrimination in renting or buying while large-sized families may be unable to find suitable living quarters (in size). Finding adequate replacement housing may be particularly troublesome for elderly households who often have low and more or less fixed incomes and few prospects for economic upward mobility. To this, one can add the special locational needs of the elderly. Many inner-city areas populated by the elderly contain senior citizen centers or groups which provide them with special services. Moreover, the elderly tend to live in areas which are easily accessible to shopping, medical and other necessary basic services. To the extent that displaced elderly individuals cannot relocate in areas equally close to these services, displacement will particularly affect them. The elderly may also suffer disproportionately through the loss of old social ties and familiar surroundings.

Lower-income ethnic and racial groups are also likely to be disproportionately affected by the loss of social and community ties through forced relocation. The political influence which these groups have been able to accrue as a result of their geographic concentration may be threatened by a massive relocation from their traditional neighborhoods. Minorities are also likely to confront the most serious difficulties in attempting to penetrate the discriminatory, exclusionary barriers erected by suburban communities.

New neighborhood residents stand to gain considerably from revitalization. Most likely, they are able to obtain lower-cost housing than they would in other parts of the metropolitan area. They rapidly accrue homeowner equity as prices go up and stand to gain from increased proximity to employment and city amenities and from lower transportation costs. At the same time, these individuals take calculated risks and may incur both psychic and material costs. Adjustment to central-city living may be difficult and oftentimes dangerous. Moreover, if early inmovers do not generate a stream of followers and the neighborhood does not revitalize, considerable equity losses can be incurred. Those who are parents may also face the costs of either lower-quality education for their children or higher tuition rates for private schooling.

Looking at revitalization from the perspective of different *neighborhoods,* most of the benefits are likely to accrue to the revitalizing neighborhoods while most of the costs are likely to accrue to the recipient neighborhoods—be they other city neighborhoods, suburban municipalities, or unincorporated areas. Revitalizing neighborhoods are likely to be better off for a variety of reasons, i.e., the upgrading in the housing stock, the increase in the purchase of rehabilitation services and materials, and the creation of renovation jobs (although it's not certain that these would accrue to the revitalizing neighborhood), improved municipal services and the availability of new financing. At the same time, revitalizing neighborhoods generally experience extensive social change whereby existing neighborhood networks and institutions of self-help, social control, and planning are disrupted. Racial and social class tension—even violence—may ensue. An indeterminate effect of rehabilitation for the revitalizing neighborhood is the probable loss of diversity in the characteristics of its population as the general trend has been for more homogenous upper-status individuals to replace existing residents. Whether this represents a cost or a benefit is dependent, it would seem, on the predilections of the observer.

Recipient neighborhoods stand to lose from revitalization in terms of a possible increase in the demand for municipal services as more dependent populations relocate in these communities, and in a possible reduction of the tax base through the influx of lower-income residents. Increased racial and social class tension may also be a problem.

From the *city* perspective, revitalization is largely a benefit as increases in the tax base and in employment opportunities accompanied by decreases in the costs of services required by the low-income population foretell the long-sought-after upswing in fiscal strength and stability. There is a single entry in the central city cost column—providing improved services in revitalizing areas. In order to attract and retain middle-income households, city governments may ultimately need to increase substantially the level of services provided to the area. Although early inmovers may expect little in the way of municipal services, those who come later will probably expect much more. If revitalization is to be encouraged and sustained, the city may have to increase public services. Whether those expenditures are likely to be offset by increased tax revenues is as yet unknown. An indeterminate effect of revitalization for cities is the possible decrease in school enrollments resulting from a decline in the number of children (given the influence of childless couples and single person households). How this would affect the quality of schools is uncertain.

Looking at revitalization from a macroperspective—that of the *metropolitan area as a whole*—the costs and benefits probably cancel each other out. On the positive side, metropolitan areas stand to gain from the revitalization of their central cities and from a possible redress of socioeconomic disparities between central cities and suburbs. On the negative side, those problems which have most plagued central cities—dependent poverty populations—are not at all remedied by revitalization. Instead, the problems of poverty, unemployment, and despair are most likely only shifted to other parts of a metropolitan area. This shift is likely to be problematic as some communities will no doubt prevent displacees from entering through exclusionary and no-growth barriers, while other communities will suffer from the increased load on their services and from heightened racial and social tension. The overall result of revitalization for metropolitan areas may only be a new pattern of metropolitan racial and economic resegregation.

From the perspective of the *nation,* neighborhood revitalization holds out the prospect of regenerating the nation's central cities and, in the process, of helping to conserve existing capital investment in housing and neighborhood infrastructure. The revival of inner city neighborhoods might also aid energy conservation by reducing transportation needs and promoting the conservation of suburban land by funneling new housing demand into the existing stock, not new construction. Another potential benefit might be the restoration of local fiscal balance, particularly since the federal government is likely to be pressed to provide loans and grants to local governments whose revenues are declining while public service demands increase.

At the same time, contrary to the claims made on its behalf, revitalization is no panacea to the problems of urban America. Revitalization does nothing to alleviate the problems of poverty and probably only works, as suggested above, to shift these problems to other parts of a metropolitan area, resulting in a new pattern of metropolitan racial and economic segregation. What national costs and benefits these new patterns may bring forth is difficult to foretell at this point. Additional potential national costs are the demand for increased housing subsidies targeted specifically to families who wish to remain in regenerating areas and relocation assistance for those who are forced to move from the area due to eviction by private owners or speculators.

The chart highlights two major questions related to revitalization which should become the subject of public policy concern: (1) to what extent are the displaced owners and renters, particularly those who are poor or who fall into the demographic, ethnic, and racial categories alluded to earlier better or worse off in their new locations? and (2) to what extent are the problems of the city being shifted to other parts of the metropolitan area, and with what consequences? The following section summarizes the very limited data available on these questions.

DISPLACEMENT: THE EMPIRICAL EVIDENCE

Displacement is the most difficult aspect of revitalization to examine systematically, partly because it is exceedingly difficult to research. A convenient definition of displacement is the one provided by James (1977a): "displacement occurs when the housing choices of people are so restricted by private market forces or public policy that they are forced to move either to another house or to another neighborhood." The essence of this definition is the notion of *involuntary* movement, i.e., *being forced* to move. The problem with this is that it requires ascertaining from those who are moving out of neighborhoods undergoing revitalization the extent to which their moving is voluntary or involuntary, and such data are not yet available. Displacement may also precede the reentry of new residents by a significant period of time, thus making it very difficult to identify at what point in the neighborhood renovation process residents are or are not being displaced.

Displacement as a result of neighborhood revitalization also needs to be placed within the wider context of residential moves in general and of other forms of displacement. Americans traditionally are a mobile people. Downs and Lachmann (1977) estimate that about 20% of the population moves every year while the mobility rate for renters is even higher (38% are estimated to move yearly). The attempt to differentiate displacement moves from these "natural" or "expected" mobility rates is further exacerbated by a number of factors.

It is very difficult to ascertain whether residents of neighborhoods undergoing revitalization would wish to move from central city locations to potential locations elsewhere in the metropolitan area, such as the suburbs. There are very few empirical data on this question. Suggesting a positive reply, for example, an Urban League study of residents being displaced out of the Capitol Hill area in Washington, DC found that even though 74% of the respondents indicated that they "felt a part of (the) neighborhood," two-thirds stated that they would like to move out of their neighborhoods to better ones if they could afford to (Washington Urban League, 1976). Similarly, in his study of Bay Village and West Cambridge in Boston, Pattison (1977) reported that the residents of the area, by and large, viewed revitalization positively as an opportunity to fulfill a long-term goal of leaving the neighborhood and moving to the suburbs.

On the other hand, one should take note of the many barriers which hinder the mobility of low-income individuals out of central-city neighborhoods and which tend to suggest that much of the moving is probably forced. Some barriers are internal to the people themselves, i.e., they are: (1) economic (lack of financial resources); (2) psychological (culture of poverty, lack of future orientation, inability to perceive alternative options, and lack of familiarity with other geographical locations); (3) social (break-up of important family, social, and neighborhood ties); and (4) political (loss of political power base in central city areas). Other barriers are external to the residents and represent societal forces which potentially hinder outward movement. Among these are: (1) the limited growth policies of local governments which prevent low-income individuals from settling in particular communities; (2) the public's generally negative attitudes toward low-cost and integrated housing; and (3) past housing policies which have contributed to the creation of metropolitan segregation (Fishman, 1978).

To complicate matters further, it is also necessary to locate the displacement that results from neighborhood revitalization within the context of other forms of displacement. As the Griers (1978) point out, there are multiple publicly and privately induced sources of displacement, such as abandonment, arson, code enforcement, highway construction, historic area designation, urban renewal, and so on. There are few reliable estimates of the extent of all forms of displacement. Meek summarizes the only available national data base on this question—HUD's Annual Housing Survey (AHS):

> According to AHS estimates, an average of approximately 500,000 U.S. households are displaced each year. Fourteen percent are displaced because of government action; the remaining 86% because of private activity. Between 1973 and 1976 the AHS estimated that more than two million households were displaced. Moreover, since the AHS relies on extremely small samples and is admittedly not very accurate when extrapolated to the national level, it may underestimate displacement, particularly that caused by government action [Meek, 1978: 3].

The proportion of displacement moves which are due to neighborhood revitalization (in comparison to other sources of displacement) is impossible to pinpoint at present. Some reports (Holman, 1977; James, 1977a, b) point to neighborhoods where wholesale shifts of population have occurred as proof of the seriousness of the problem. Others (Pattison, 1977) can point to neighborhoods such as West Cambridge where renovation and replacement of the existing population by newcomers seems to have occurred through a process of "natural attrition." A recent study of fourteen selected cities by the Griers (1978) suggests that displacement due to private renovation may be a smaller problem than is commonly believed. Although their reconnaissance of selected cities was not based on scientific sampling techniques, they estimate that "in most cities the numbers displaced because of revitalization are no higher than the low hundreds, and probably do not exceed the low thousands even in the most active, like Washington, D.C." The authors go on to note that, "The total displacement problem, however, appears to be a large and rapidly-growing one. Reinvestment-related displacement is important to treat because its high visibility makes it a target for emotions triggered by the larger problem; but it should not be mistaken for that problem" (Grier and Grier, 1978: iii). The root causes of displacement, in the Griers' view, are changes in the private market. Displacement "is stimulated by broad market forces which are eroding the supply of housing at low-and-moderate-price levels. It is aided by public action or inaction which tends to force displacement into the private sector, where it is harder both to detect and to remedy" (Grier and Grier, 1978: iv).

The intensity of displacement problems deriving from revitalization seems to vary according to the type of neighborhood and according to different stages in the revitalization process. In Boston, Chicago, and Cleveland, much of the rehabilitation has been targeted toward existing owner-occupants of older houses. Pattison (1977) reports a similar finding in his West Cambridge case:

> Displacement, i.e. where long-term residents feel a pressure to move, does not seem to have been a characteristic of the gentrification process in West Cambridge. What has occurred is that incumbents have not been replaced (as they leave for whatever reason) by people of similar means and occupations.

On the other hand, in neighborhoods where there are a large number of multifamily dwellings and where the renovation pace is rapid, displacement is a problem—as in Washington, DC.

Displacement may also affect different types of residents at different stages in the revitalization process. In many cases, displacement of old-time residents may precede the entry of new residents. First to be affected are the renters. As Schur suggests, using New York City as an example (1977), as real estate developers and speculators buy up properties in the soon-to-be revitalized neighborhoods, they have no incentive to retain the present tenants (even on an interim basis) even though actual renovation and occupancy by the new brand of immigrants may be several years away. This is so for a variety of reasons: (1) rents in these neighborhoods are so low that current costs of maintenance and operation are not being covered; (2) where housing code violations exist and maintenance fails to keep pace with on-going deterioration, tenants complain or take matters in their own hands by withholding rent (thus becoming a nuisance to the property owners); and (3) finally, when the time for sale comes, the property is worth more vacant than when occupied.

As the neighborhood revitalization process develops and as the pioneers are joined by additional affluent home-buyers attracted by the early activity, a second stage of displacement begins. This time those affected are the owner-occupant long-time residents. Cognizant of seemingly good offers, they may sell too eagerly and too rapidly thinking that the neighborhood is still in decline. By not realizing the true value of their property, the former neighborhood homeowner may find it difficult to purchase housing other than in areas similar to those left behind. A third stage of displacement occurs when the neighborhood has been fully revitalized. As assessments on neighborhood property increase, many long-time owner-occupants may be driven out because of the rise in property taxes. At this stage, it is the elderly on fixed incomes who are particularly affected (Kollias 1977).

Who is being displaced? A recent HUD report on displacement prepared for the Congress (U.S. Department of Housing and Urban Development, 1979b: 24) notes that: "Despite the paucity of available data, certain themes are relatively consistent. Elderly households, minority households, and renters are repeatedly reported to be among the displaced. Working-class and blue-collar households have also been identified as frequent out-movers." A study of Capitol Hill in Washington, DC, for example, supports this general conclusion. Out of sixty-four identified displacees, thirty were families with three or more children and most were black. The remainder included a large portion of elderly families (James, 1977a).[2] Another study (Clay, 1979), found an exodus of working-class and blue-collar households from neighborhoods experiencing displacement. While 95% of the revitalizing neighborhoods were dominated by these groups prior to revitalization, only 26% still contained significant numbers of such households at the time of the study. In a study of persons who moved from apartment buildings undergoing condominium conversion in Washington, DC, 45% of the out-mover households were found to be elderly, 82% contained one or two persons and over two-thirds had incomes (in 1975) of less than $15,000 (Development Economics Group, 1976). Slightly over half of these households responded that they chose to move rather than to purchase their apartments because it would have been too expensive to stay.

Where do displacees go? The very limited evidence that is available on the destination of displacees suggests two things: (1) most people move very short distances, as close as possible to the original location; (2) most people move several times. The data available on these questions are summarized below.

The Urban Institute traced the destination of families displaced in three subsidized housing programs in Baltimore. Major findings were that a large proportion moved within a mile of their old homes and that many who moved out of the area moved back when they could (Stanfield, 1977). A study of displacees from Seaton St. and Ontario Road in Adams-Morgan in Washington, DC showed that the majority of individuals had moved to the immediate surrounding area; they were paying higher prices for less space; a number of them had moved up to three times since being displaced from the original location; the few that had left the area altogether tended to come back on the weekends and evenings to partake in religious and social gatherings (Smith, 1977a,b).

A study of people being displaced out of the Capitol Hill area in Washington, DC showed that one-half of the displacees relocated within Capitol Hill to apartments where they remained threatened by dislocation; 40% relocated to other row house neighborhoods in DC, most of these in Adams-Morgan, and only three out of twenty-nine families left the District—two of these went to Prince Georges County in suburban Maryland (James, 1977a). The Urban Coalition study (Holman, 1977) reports similar findings; displacees moved "to the nearest place they could." In addition, there is some evidence that indicates that people currently being displaced out of neighborhoods undergoing revitalization are, in some instances (e.g., Queen Village in Philadelphia, Mount Auburn in Cincinnati) former displacees from urban renewal projects of the 1950s and 1960s (Weiler, 1977). Finally, similar but more optimistic findings are suggested by a study of condominium conversion in Washington, DC conducted in 1976. Nearly 90% of the families remained in the city and over three-quarters of this group relocated in the same or adjacent neighborhoods. Although the households affected by condominium conversions are not typical of all households threatened by displacement, it is instructive to note that these families fared fairly well in their search for replacement housing; by and large they were able to locate homes of similar cost and size in areas near those they left and over two-thirds were able to find a new home within one month of beginning their search (Development Economics Group, 1976).

There are other cases which suggest that forced moves do not necessarily result from neighborhood revitalization. Commenting on the experience in Bay Village and West Cambridge in Boston, Pattison (1977) observes:[3]

In Bay Village displacement of homeowners does not appear to have been a major characteristic of the upgrading process. No one cited rising property taxes or a sense of alienation from newcomers as forcing out long-term residents. One realtor observed that people were vacating their properties because "they were through with them. . . . Many folks were just dying off." And there were those, an ex-resident recalled, "who really wanted out. . . . [They] had lived in the neighborhood through all the bad times and decided that it couldn't get better . . ." and they stuck with their attitudes—the visible signs of upgrading notwithstanding. Instead, upgrading and the accompanying increased demand provided some incumbents with the financial opportunity to fulfill a long-term goal—to leave the neighborhood and move to the suburbs.

CONCLUSION

While revitalization and its attendant consequences do not seem, at present, to represent a quantitatively significant phenomenon in terms of the total number of households affected, the movement is sufficiently complex and ambiguous to warrant focusing concerted public attention at both national and local levels on this question. Displacement problems currently in evidence need to be dealt with and the potential metropolitan consequences of revitalization need to be considered and planned for. While specific policies which might be considered are beyond the scope of this chapter (see, for example, the discussion in Grier and Grier, 1978), it is clear that a true national-local partnership needs to be forged on this issue. As James (1977a) has put it, displacement is a national issue and a local problem. Local governments and local community groups combating displacement are in need of federal assistance, know-how, and, possibly above all, of moral leadership. Such moral leadership has not been readily forthcoming as many in Washington—enamored by the vision of a "revitalized urban America"—have been lured by the presumed fiscal advantages of revitalization and have not paused to consider the broader implications of this phenomenon nor the potential distribution of its costs and benefits.

NOTES

1. The Urban Coalition survey relied on a mail survey technique. Local officials in cities where the coalition had local affiliates, associates, or other contacts were asked to supply information on the extent of private renovation in their cities, including the number, type, and location of houses being rehabilitated and the characteristics of the renovators. Despite the important insights provided by this survey, it suffers from possible problems of unrepresentativeness. There is no assur-

ance, moreover, that the responses are consistent or reliable; the extent of rehabilitation reported by various respondents likely ranged from off-the-cuff estimates to hard data taken from systematic searches of building permits or other files.

2. One has to be careful, however, in interpreting these data. They are based on the records maintained by Friendship House, a neighborhood service organization. These households are not likely to be representative of all the families who left Capitol Hill during renovation. Rather, they indicate the types of families most severely affected by displacement and families who sought some form of assistance in relocating.

3. Care must be taken in interpreting Pattison's data. Pattison's methodology consisted of interviewing households whose names were obtained from various "leads." These respondents were then asked to suggest the names of other residents who would be amenable to interviews. As a result, respondents tended to be people who were active in the local neighborhood association, which suggests possible bias in Pattison's results.

REFERENCES

BLACK, J. T. (1975) "Private market renovation in central cities: a U.L.I. survey." Urban Land 34 (November): 3-9.

Boston Globe (1977) "Pressures on the North End." (June 11).

CLAY, P. (1979) The Neighborhood Renewal Game: Middle Class Resettlement and Incumbent Upgrading in the 1970's. Lexington, MA: Lexington Books.

Development Economics Group (1976) Condominiums in the District of Columbia. Report to the Office of Housing and Community Development, Government of the District of Columbia.

DOWNS, A. and J. L. LACHMANN (1977) "The role of neighborhoods in the mature metropolis." Prepared for symposium on challenges and opportunities in the mature metropolis, St. Louis, June 6-8.

EMBRY, R. (1977) Testimony presented before the U.S. Senate Committee on Banking, Housing and Urban Affairs, Washington, DC, July 7.

FISHMAN, R. (1977) Housing for All Under the Law: New Directions in Housing, Land Use and Planning Law. Cambridge, MA: Ballinger.

GALE, D. E. (1976) "The back-to-the-city movement, or is it?: a survey of recent homeowners in the Mount Pleasant neighborhood of Washington, D.C." Washington, DC: George Washington University Department of Urban and Regional Planning.

——— (1977) "The back-to-the-city movement revisited: a survey of recent homebuyers in the Capitol Hill neighborhood of Washington, D.C." Washington, DC: George Washington University Department of Urban and Regional Planning.

GRIER, G. and E. S. GRIER (1977) Movers to the City: New Data on the Housing Market for Washington, D.C. Washington, DC: Washington Center for Metropolitan Studies (May).

——— (1978) "Urban displacement: a reconnaissance." Memo report prepared for the U.S. Department of Housing and Urban Development (March).

HARTNETT, K. (1977) "Tracking the return of the gentry: the bad side of central-city chic." Boston Globe (May 28).

HOLMAN, C. (1977) Testimony presented before the U.S. Senate Committee on Banking, Housing and Urban Affairs, Washington, DC, July 8.

JAMES, F. J. (1977a) Back to the City: An Appraisal of Housing Reinvestment and Population Change in Urban America. Washington, DC: Urban Institute.
————— (1977b) "Private reinvestment in older housing and older neighborhoods: recent trends and forces." Testimony presented before the U.S. Senate Committee on Banking, Housing and Urban Affairs, Washington, DC, July 8.
KOLLIAS, K. (1977) Internal memoranda on displacement. Washington, DC: U.S. Department of Housing and Urban Development, September to December. (unpublished memoranda)
KORNEGAY, S. (1977) "Urban residents should reconsider moving, prof. says." Washington Post (November 12).
LINEBERRY, R. and I. SHARKANSKY (1974) Urban Politics and Public Policy. New York: Harper & Row.
LIPTON, G. (1977) "Evidence of central city revival." American Institute of Planners Journal 43 (April): 136-147.
MEEK, C. (1978) "Public displacement from 1952-1977." Working paper, Office of Community Planning and Development, U.S. Department of Housing and Urban Development.
PATTISON, T. (1977) "The process of neighborhood upgrading and gentification." Unpublished master's thesis, Department of City Planning, Massachusetts Institute of Technology.
REINHOLD, R. (1977) "Middle-class return displaces some urban poor." New York Times (June 5).
ROGG, N. H. (1977) Urban Housing Rehabilitation in the United States. Washington, DC: United States League of Savings Associations (December).
ROSS, N. L. (1977) "Encouraging renewal." Washington Post (November 12).
SCHUR, R. (1977) Testimony presented before the U.S. Senate Committee on Banking, Housing and Urban Affairs, July 7.
SMITH, F. (1977a) "Displacement in Adams-Morgan." Testimony presented before the U.S. Senate Committee on Banking, Housing and Urban Affairs, July 7.
————— (1977b) Personal interview, Washington, DC, December 9.
STANFIELD, R. (1977) "The human renewal." National Journal 9 (February 19): 290.
STERNLIEB, G. and J. W. HUGHES (1977) "New regional and metropolitan realities of America." American Institute of Planners Journal 43 (July): 227-241.
U.S. Department of Housing and Urban Development (1979a) Office of Community Planning and Development, Urban Policy Staff. "Whither or whether urban distress—a response to the article 'The urban crisis leaves town,' by T. D. Allman, Harper's, December, 1978." Washington, DC. (mimeo)
————— (1979b) Displacement Report to the U.S. Congress. Washington, DC (February).
Washington Urban League (1976) SOS 1976: Speakout for Survival. Washington, DC (June).
WEILER, C. (1977) Testimony presented before the U.S. Senate Committee on Banking, Housing and Urban Affairs, Washington, DC, July 8.

3

Housing, Lending Institutions and Public Policy

RICHARD HULA

☐ THE DECLINING ECONOMIC VIABILITY of older urban areas is well documented, particularly the migration of industry and jobs to the suburbs or perhaps completely out of the region. This period of economic transition is also associated with demographic change with the cities increasingly becoming the home of low-income minorities.

Implicit (and perhaps explicit) in this particular vision of urban development is that the observable decline in many American cities represents a natural process. An early example of this approach is found in the work of Robert Park and Ernest Burgess. Park and Burgess tried to explain neighborhood transition and change through the use of a number of terms and concepts borrowed from biological ecology. Thus, they spoke of "neighborhood succession" and "dominant land use."[1] Although the language is often quite different, a similar view often permeates urban studies in a variety of social science disciplines. Obviously, this view has significant policy implications. It would seem to restrict public intervention into any urban crisis to the amelioration of the conditions of those who are most affected by the inevitable transformation to new urban forms. Any attempt to alter the process of urban development seems certain to fail. Indeed, numerous observers argue that the decline of central cities carries a number of benefits for those who remain. In the area of housing, for

AUTHOR'S NOTE: *This research has been supported by a grant from the University of Texas at Dallas. Special thanks are in order to Danny Nuckols, Carla Duchin, and Jarol Krause who aided in the collection of the data. Thanks also to Alex Stepick, Martin T. Katzman, and Donald Rosenthal for helpful comments. An earlier version of this paper was presented at the 1979 Meetings of the Midwest Political Science Association.*

example, it has long been argued that there exists a filtering process by which housing of relatively high quality is left to low-income populations as the more affluent leave the central cities.[2]

More recently, a number of scholars and policy-makers have come to challenge the inevitability of much of this urban decline. A number of governmental policies have been offered in support of this view. For example, there is a good deal of evidence that the traditional emphasis of the FHA on large suburban (all-white) housing tracts has contributed to the widespread "demand" for such suburban housing. Another common example is the widespread use of zoning regulations by local governments, which greatly restricts the ability of low-income populations to migrate toward suburban-based jobs. This latter example is particularly ironic in that migration toward jobs is often cited as a basic factor in the "natural" development of the earlier industrial city. As it has become clear that public intervention has had some impact on central city decline, a number of aspects of the private market have also been subject to scrutiny. This chapter will examine the general role of private lending institutions within the housing market of a large urban county (Dallas, Texas). Specifically, it explores whether decisions which restrict lending commitments to central city areas can be explained by "natural market forces." Various policy alternatives for increasing credit to inner city areas will also be considered.

THE ROLE OF LENDING INSTITUTIONS
IN THE PROCESS OF URBAN DECLINE

It is obvious that lending institutions play a fundamental role in the housing market.[3] Few individuals are able to purchase homes outright and most, therefore, rely on institutional sources of financing. This is also true for major home repairs or improvements. The availability of financing is also of critical concern to the real estate investor for whom the opportunity to refinance property represents a major source of capital and improvement funds. A lack of home credit makes it difficult (at best) to sell property. It also results in an almost inevitable physical decline of property as needed maintenance is deferred. This decline may set in motion a vicious cycle which can lead to the abandonment of apparently sound housing.

Unlike many issues of public policy, a number of "facts" which describe the allocation of home credit are largely accepted by both critics and defenders of lending institutions. There exists a good deal of empirical work from which a number of generalizations can be drawn. Five such generalizations are suggested below:

(1) There are relatively few home mortgage or home improvement loans originated in most older sections of American central cities.[4]

(2) When mortgage loans are originated in older urban neighborhoods, they tend to be underwritten by the federal government through the FHA or VA programs (Lyons, 1975).

(3) Contrary to statutory intent, federally guaranteed loans usually involve increased costs to the borrower when compared to conventional financing (Bradford, 1979; Harvey, 1973).

(4) There exists an empirical association between the physical and economic decline of urban neighborhoods and a lack of conventional credit resources.[5]

(5) There exists an empirical association between changing racial makeup of neighborhoods (i.e., the arrival of minorities) and a lack of conventional credit resources.[6]

Although there is a measure of agreement on these descriptive generalizations, there is no corresponding consensus on what they mean. A number of observers have, for example, argued that the empirical association between a lack of conventional mortgage credit and neighborhood deterioration implies that lending institutions, at least in part, are the cause of this deterioration. Lending institutions in a number of American cities have been subject to vociferous attack by community organizations which claim that the institutional policy of not making loans in their neighborhoods has led to both community blight and a process of capital exportation in which local savings deposits have been used to support the development of other sections of the city (Bradford and Rubinowitz, 1975).

The specific mechanism by which lending institutions are thought to bring on neighborhood deterioration is *redlining*. In its purest form, redlining is the refusal of a lending institution to grant any loan in a particular geographic area. That is, if an area is redlined no applicant, however credit-worthy, will be granted a loan. Recently, somewhat more sophisticated definitions of redlining have included not only the outright refusal to make loans, but also the existence of differential terms applied to all loans within certain areas. For example, savings institutions may require relatively greater down payments, shorter repayment periods, or insist on some form of mortgage insurance. Each of these conditions raises the cost of the loan to the consumer, and can be thought of as making the loan less attractive. Many argue that such differential loan conditions can have much the same effect on a neighborhood as a decision to make no loans at all.

Closely tied to the notion of redlining is the issue of neighborhood disinvestment. Many who claim that savings institutions engage in redlining argue there is no similar aversion to accepting deposits in allegedly high-risk communities. Thus, there occurs an exportation of capital to other areas within the city. Critics of lending institutions often cite the existence of relatively unfavorable ratios of deposits to loans to "prove" that there is sufficient financial resources within redlined areas to warrant institutional investment.[7]

Those who are more sympathetic to the lending industry respond that the withdrawal of home credit to a neighborhood is not the cause of deterioration, but rather a reaction to it. Lenders claim that they simply follow normal market criteria. Indeed, they are forbidden to do otherwise by numerous federal and state supervisory agencies charged with regulating them. For example, while there is language in Section 801 (a) (3) of the *Housing and Community Development Act, 1977* which states that savings institutions have a responsibility to the communities in which they are located, it is also explicit in requiring that loan policies follow sound lending practices. Lenders see the issue as a question of institutional responsibility to savers and shareholders. Often charges of redlining are seen as attacks on the entire market system rather than on the behavior of specific lending institutions. After dismissing the claims of numerous studies which purport to show redlining exists in a number of American cities, George Benston concludes:

> Perhaps the best explanation of the anti-redlining crusade is that it ultimately reflects not a belief that banks are really discriminating, but a profound suspicion of the market system. What the crusaders really want is to allocate credit—to have government, and not the market—make decisions about who gets mortgage money. The belief that the market cannot be trusted where housing is concerned is nothing new, of course. It is already profoundly represented in the mind-boggling array of subsidies that are now administered by the U.S. Department of Housing and Urban Development" [Benston, 1978: 69].

A SHORT NOTE ON THE LEGAL ISSUES

A review of the various charter regulations suggests that of the major components of the mortgage market, only savings and loan associations seem to be under an apparent legal obligation to be responsive to the immediate geographic territory in which they operate, Darel E. Grothaus (1975), reviewing the regulation of federally chartered savings and loan associations, stresses the importance of the "service area" in the review of applications for new savings and loans. A similar geographic responsibility seems clearly implied in the Texas *Laws and Regulations for Savings and Loan Associations.* Statute 2.08, which defines criteria for approving charter applications, states that the applicant must show that "there is a public need for the proposed association and the volume of the business *in the community in which the proposed association will conduct its business* is such to indicate profitable operation" (Texas Stat. 2.08 (3); emphasis supplied).

Although there is a formal requirement that both state and federally chartered savings and loan associations respond to local credit needs, the substantive significance of these regulations is not clear. Foremost is the problem of defining local credit needs, particularly some time after a charter application has been approved. Savings and loan associations can (and often do) explain differential loan rates across geographic areas by variations in demand as well as the availability of "sound" loans within a given section of a city. Reinforcing the ambiguity surrounding these requirements of local responsibility are regulations which demand that savings and loan associations commit loans in a "sound" manner. There seems little doubt that where there might be a conflict between sound fiscal policy and local responsibility, sound fiscal responsibility is stressed by both federal and state regulators.

A more successful legal objection to redlining has been based on general statutes which apply to all lending institutions. This attack is based on the premise that redlined neighborhoods usually have relatively greater concentrations of minority populations than neighborhoods in which home credit is more available. Thus, the act of redlining may have as a consequence the reduction of home credit to minorities. If this is the case, it may be illegal under the terms of federal fair housing and civil rights laws. Where it has been possible to show that the aggregate credit policies of lending institutions have had a differential impact on populations (rather than geographic areas) the courts have shown a willingness at least to entertain the notion that violations of federal law may exist.[8]

While the legal aspects of redlining are certainly important, it does seem that the nearly exclusive focus on legal questions ignores other policy issues. Specifically, this chapter is interested in the impact of private sector, i.e., lending institutions', policy in the decline of inner city neighborhoods. In a sense, the policy issues are broader than whether a discrimination case can be made against a set of lending institutions. To the extent that it can be shown that aggregate lending policies lead to neighborhood decline, it may be possible to formulate public policy to reduce such impacts.

OUTLINE

While it is certainly possible (indeed useful) to consider the role of lending institutions in the development of urban areas at an abstract level, it is probably impossible to understand the dynamics of the home credit market outside the context of an actual market. Thus we turn to a case study of a major component of the home credit market operating in Dallas County (Texas) for the year 1977. We wish to explore the distribution of all federally insured loans, conventional mortgages and home improvement loans originated or purchased by savings and loan associations operating in the county. Specific empirical questions to be addressed include:

(1) What is the geographic distribution of home credit on the part of county savings and loan associations?
(2) Is there a consistent pattern of allocation for different types of home credit?
(3) Is the allocation of home credit consistent across different savings and loan associations?
(4) What are the characteristics of high- versus low-loan neighborhoods?
(5) What is the relationship between deposits and the allocation of home credit?

The relatively rapid development of the northern sections of Dallas County has had important implications for the city of Dallas, since in general the most rapid growth has occurred outside the corporate boundaries of the city. By contrast, the southern sections of the county are within the city. The resulting demography of the county is a familiar one to students of "urban decline." Over the past few years the proportion of middle-class (often white) residents of the central city has declined.

THE DATA

The federal *Home Mortgage Disclosure Act of 1975* requires that all depository lending institutions make a public disclosure of where certain categories of home loans are made. Specifically, Regulation C of the Federal Reserve which implemented the Act requires that lending institutions disclose by census tract the number and total value of federally insured home mortgages, conventional mortgages, home improvement loans, loans for multifamily dwellings and home mortgages for nonresident owners. Data have been collected from twenty-five of twenty-eight savings and loans which operate in the county. One institution has less than ten

million dollars in assets and is thus exempt from the reporting requirements. Two other institutions have refused to provide the information as
required by federal law.

Although the data are reported by census tract, much of the analysis
reported in this paper will be based on the somewhat larger geographic
units of statistical communities. Dallas County has been divided into some
fifty communities by the city of Dallas planning department. Particular
communities are defined by a number of contiguous census tracts. For
areas outside the city of Dallas the communities generally take in a
suburban municipality. Within the city itself there has been some effort to
specify relatively homogeneous neighborhoods. The utilization of such
communities has a number of benefits both practical and conceptual.
Practically, it reduces the number of cases from almost three hundred (the
number of census tracts in the county) to fifty. For the communities
within the city it also allows us to use a good deal of information which
has been collected by the city through a number of social surveys. More
importantly, the larger geographic unit of statistical community may be
more appropriate for our examination of lending institution impacts on
neighborhoods. The census tract we suspect is simply too small.

Before we proceed to the actual analysis, several limitations of these
data should be discussed. Foremost is the limited coverage of the *Home
Mortgage Disclosure Act*. The legislation requires only deposit institutions
to report where home loans are originated or purchased. Excluded from
the act are mortgage bankers who are a major source of federally insured
loans. Table 3.1 gives some measure of the importance of this exclusion.[9]
We see that for a sample of all real estate transactions for the year 1976
some 56% are FHA/VA loans. A second exclusion in these data is lending
institutions which operate in Dallas County, but do not actually have

TABLE 3.1   Summary of the Number and Value of Home Mortgages in
Dallas County by Type, 1976[a]

Mortgage Type	Number	Total Value	Average Value	Proportion of Total Number	Value
Conventional	955	$49,733,200	$52,076	44%	61%
FHA	868	21,506,606	24,777	40	26
VA	333	8,852,900	26,585	15	11
Other[b]	27	1,114,500	31,277	1	1

a. These data are derived from a 20% sample of all real estate transactions for Dallas
County for the year 1976. The data are taken from reports published by the area
American Association of Real Estate Appraisers.
b. Includes cash, contract, and other forms of home sales.

branches physically within the county. While these institutions certainly report their loans, there seems to be no practical way to discover who they are. At present, we have been forced to assume that the magnitude of these loans is not significant.

There is a good deal of evidence that there may be significant error in the data as the result of carelessness on the part of collecting institutions. A recurring problem in the coding of the data is the reporting of nonexistent census tracts. For some institutions these tracts may make up a full 5% of the total number of the reported loans. Given this fact, it seems reasonable to assume that there is a similar error within those tracts which at least exist. Neither the reporting institutions nor the appropriate regulatory agencies have shown much interest in this particular problem.[10] We are forced to accept this error and simply assume that it does not introduce systematic bias into our analysis.

THE ALLOCATION OF HOME CREDIT IN DALLAS COUNTY, 1977

An inspection of the empirical distribution of home credit reveals that for all types of loans there is a high degree of concentration in relatively few statistical communities.[11] For example, five communities account for 41% of all conventional mortgages. A slightly different set of five communities accounts for 51% of all FHA/VA mortgages. A third set of five communities account for 37% of all home improvement loans. By way of contrast, ten communities have a combined total of less than 1% of conventional mortgages, FHA/VA mortgages and home improvement loans. Also apparent in the distribution is the consistency with which the newer, generally more affluent, areas of North Dallas County receive the bulk of home credit. This geographic concentration is summarized in Table 3.2, which reports the number and total value for various loan types by geographic area. The county zones used in Table 3.2 have been aggregated from statistical community data, and were constructed so as to roughly divide the county into four equal areas. The county's northern and southern borders act as two zone boundaries, the second two are imaginary east-west lines. Further support for the notion of a consistent pattern of credit allocation is provided by the positive correlation between various types of home credit. Only the association between number and value of FHA/VA loans and other forms of credit is weak. Inspections of the scatterplots between various forms of home credit (these plots are not reproduced here) suggest that FHA/VA loans are most common in tracts with a moderate number of conventional mortgages and home improvement loans. FHA/VA loans are more rare in tracts which have either many or very few loans.

TABLE 3.2 Summary Geographical Distribution of Home Credit in
Dallas County, 1977

| | County Zone | | | |
Loan Category	Far North	Near North	Near South	Far South
Number FHA/VA	101	68	22	8
Value of FHA/VA	$ 3,459,000	$ 2,218,195	$ 566,694	$ 250,575
Number of conventional mortgages	7265	3831	859	411
Value of conventional mortgages	$361,471,805	$155,062,849	$24,152,834	$15,008,669
Number of home improvement loans	996	520	224	26
Value of home improvement loans	$ 10,717,961	$ 5,130,523	$ 1,540,493	$ 200,575

This clear geographic pattern in Dallas County could be the result of a number of circumstances. It may be that the largest savings and loan associations are most active in the northern areas of the county and thus, at least in aggregate terms, overwhelm the loan policies of the smaller associations. Certainly this is possible since the three largest savings and loans operating in the county hold approximately 66% of the total county deposits. It may also be the case that this pattern reflects lending decisions throughout the county. To determine which of these alternatives is correct, each statistical community was examined to see whether any savings and loan association reported at least 5% of its conventional mortgage portfolio (purchased or originated in 1977) as being within that community. A similar examination was undertaken for FHA/VA and home improvement loans. This analysis provided clear evidence that the concentration of conventional mortgages and home improvement loans is not an artifact of the lending policies of a few very large institutions. For example, sixteen of twenty-five reporting savings and loan associations report that at least 5% of their conventional mortgage loans obtained in 1977 were located in the statistical community with the greatest overall number of conventional mortgages in the county. A similar pattern is evident for home improvement loans. The case of federally insured loans is much less clear. This is due primarily to the relatively small number of such loans which are originated or purchased by institutions within the county. Indeed, only three institutions account for almost 88% of the federally insured loans.

THE DISINVESTMENT ISSUE

A key component of the redlining debate is the notion of neighborhood disinvestment. As defined earlier, disinvestment occurs when lending institutions are not willing to commit the same amount of home credit per unit of savings in all neighborhoods in which they accept deposits. It has been claimed that disinvestment represents a type of economic imperialism in which the growth and development of one area of a city are based on capital exported from another area. As with the study of redlining itself, the study of disinvestment has in the past been greatly obscured by a lack of data. In this chapter we are forced to rely on deposit information published by the Federal Home Loan Bank Board (FHLBB), which reports only the location and total savings deposits of each branch and home offices of savings and loan associations operating within Dallas County. It does not, however, indicate the source of the deposits, i.e., where individual depositors live. If we wish to claim that our data represent neighborhood-derived savings, we need to assume that an individual will choose to place deposits in a savings and loan association which is geographically convenient to his or her home. While, in general, this assumption seems reasonable, it is clearly untenable for the significant deposits in central business district (CBD) offices. In our initial deposit analysis we have contrasted a number of simple models which make different assumptions about the source of these funds. These include:

(I)   Simple exclusion of CBD deposits. Perhaps the simplest assumption one might make is that the CBD deposits are irrelevant to the neighborhood investment issue and should be dropped from the analysis.

(II) Equal allocation of CBD deposits. In this case we assume that deposits come from all portions of the county equally. Therefore, the deposits are "allocated" equally to each portion of the county.

(III) Proportional allocation of CBD deposits. Here we allocated the CBD deposits by the proportion of non-CBD deposits held in each area of the county.

Since there is general agreement that there is a good deal more wealth in the northern portions of the county, two further models which weight these areas were examined.

(IV) Equal allocation of CBD deposits to northern zones. In this case the CBD deposits were assumed to come equally from the northern portions of the county. Thus, the CBD deposits were allocated equally to the two northern zones.

(V) Allocation of CBD deposits to the far northern zone. In this final case, the far northern area of the county is assumed to be the sole source of the CBD deposits.

Even if we are able to document the source of savings deposits with some precision, there remains a serious problem in any analysis of disinvestment. One needs to specify the relevant community in which one expects a lending institution to have specific responsibility to make home credit available. How broadly ought one to define this potential target area? Large lending institutions might (and indeed do) argue that their relevant target communities are the entire metropolitan area, thus defining away any possible charge of disinvestment. Community organizations, as one would expect, tend to define target areas much more narrowly. Usually the suggested target area conforms to the organization's notion of neighborhood, even though the "social reality" of these neighborhood may be open to empirical challenge. In this chapter we will present two alternative units as possible target areas. First, we explore the distribution of home credit relative to deposits across county zones. Following this discussion, we will consider the more geographically specific statistical communities.

COUNTY ZONES

Table 3.3 reports the ratio of total loan values to total deposits of savings and loan offices operating within each of four county zones. These ratios have been calculated for conventional home mortgages, FHA/VA mortgages and home improvement loans. Perhaps the most striking feature of Table 3.3 is the extent to which deposits are concentrated in the two northern zones of the county. Almost 80% of all deposits (excluding offices in the CBD) in savings and loan associations operating in Dallas County are found in offices located in either the near or far north county zones. As we saw in Table 3.2, there was a similar concentration of home credit in these two zones. However, if we compare Table 3.2 and Table 3.3, we note a significant difference with respect to the two northern zones. Although the far northern zone has received a good deal more home credit than the near-northern zone, there were actually more deposits in savings and loan offices located in the near-northern zone. As a result, there is a rather dramatic difference between the two northern zones in the ratio of the value of home loans to the value of savings deposits. This difference is consistent no matter what assumptions one makes concerning the deposits reported for central business district offices. In Table 3.3 we find that the ratio of loan value to deposits is over 75% greater in the far

TABLE 3.3 Ratio of Value of Home Loans to Deposits by County Zone, 1977

County Zone	Total Zone Deposits[a]	Estimates of Total Loan Value to Total Deposit Ratio[b]				
		Estimate I (Excludes CBD)	Estimate II (divides CBD equally)	Estimate III (Divides CBD by proportions)	Estimate IV (Divides CBD Between Northern Zones)	Estimate V (Allocates CBD to Far North Zone)
		Conventional Mortgages				
Far North	$1,042,518	.35	.31	.29	.27	.23
Near North	1,212,692	.13	.12	.11	.10	.13
Near South	475,435	.06	.05	.05	.06	.06
Far South	34,627	.90	.18	.77	.90	.90
		FHA/VA Mortgages				
Far North	$1,042,518	.003	.003	.003	.003	.002
Near North	1,212,692	.002	.002	.002	.002	.002
Near South	475,435	.002	.001	.001	.002	.002
Far South	34,627	.009	.002	.008	.009	.009
		Home Improvement Loans				
Far North	$1,042,518	.01	.009	.009	.008	.007
Near North	1,212,692	.004	.004	.004	.003	.004
Near South	475,435	.004	.003	.003	.004	.004
Far South	34,627	.01	.002	.01	.01	.010

a. Deposits reported in units of $1,000. Source: Federal Home Loan Bank Board.
b. For description of how estimates were calculated, see text. Note: Ratios based on loans made by all savings and loan associations in sample, not simply institutions reporting deposits.

northern zone than the near north even when all the CBD deposits are assumed to be derived from depositors in the far northern zone. Table 3.3 provides strong evidence that some sort of capital "export" from the near north to the far northern zone is in fact taking place. Once again, we need to stress that the meaning of these empirical findings is by no means clear. We need to explore in some detail whether this capital export represents geographic discrimination, or reflects the normal workings of the market. We shall return to this question in a later section.

One interesting feature of Table 3.3 is the apparently high loan to deposit ratio in the far southern zone of the county. Indeed the ratio for conventional mortgages is over 200% of the ratio of the far north zone. It is difficult, however, to argue seriously that there is a significant pattern of capital exportation to the far southern zone of the county. There is simply very little activity on the part of savings and loan associations in this section of the county, be it lending or accepting savings deposits. By way of example, the southern zone reports slightly more than 1% of the county's total savings deposits (excluding CBD deposits). As a result of the relatively small amounts involved, the ratio of loan value to deposits is very unstable. For example, the ratio of conventional mortgages to total savings deposits varies in Table 3.3 from almost 90% to 18% depending on the assumptions one makes concerning the CBD savings deposits.

The ambiguity in the investment ratios for the far southern zone requires some caution in defining disinvestment as the sole issue in the redlining debate. From Table 3.3 we can see that it takes relatively few loans to achieve a very favorable deposit to home credit ratio for the far southern county zone. Nevertheless, the level of aggregate investment in this portion of the county is very small. From a public policy perspective, the balancing of deposits to loan ratios for various sections of a metropolitan area may meet a certain criteria of fairness, but will almost certainly not lead to balanced growth, much less the revitalization or redevelopment of already deteriorated areas. If there is a serious commitment to increase private investment in the far and perhaps near-southern zones of the county, some measure of aggregate investment rather than deposit to loan ratios will have to guide policy. While attempts to redirect private investment in the American economy can hardly be considered novel, in the case of redlining it does seem to redirect the argument from how best to correct imperfections in the market to consideration of at least a partial restructuring of the market itself. Such discussion, of course, generates a major fear among lenders—some form of direct credit allocation.

INVESTMENT RATIOS FOR STATISTICAL COMMUNITIES

Many would argue that the county zones used in the analysis of deposits to loan ratios reported in Table 3.3 are simply too large to capture what might be called loan target areas. To explore the investment policies of Dallas savings and loan associations in their more immediate working environment we have disaggregated loan data to the statistical community level. Using the FHLBB data, we were then able to calculate a second series of loan to deposit ratios. These ratios were computed by dividing the total deposits held in a savings and loan office by the total value of loans purchased and originated by the association within the statistical community in which the office is located. Where an association had more than one office in a statistical community, deposits for the multiple offices were added together. Thus, each savings and loan association could have only one loan to deposit ratio in a given statistical community.

There is a bewildering variance in value to loan ratios for specific savings and loan associations. For example, one association branch office committed over $500,000 in conventional mortgage credit for every $1,000 of reported savings deposits. The same association reported a ratio of one dollar of conventional mortgage credit per $1,000 deposits for a second branch. A number of associations report no loans at all in statistical communities in which they have offices. It is obvious that with respect to savings deposits some statistical communities receive proportionately less home credit than others. Indeed the difference often seems dramatic. However, one can question the substantive significance of these data. One might challenge the validity of the loan to deposit ratio measure. For example, a very high loan to deposit ratio is often a function of very low deposits as well as high institutional investment. In the very high loan to deposit ratio example given above, the office associated with that ratio is a small branch of a very large association. This suggests that a number of relatively small offices may distort the magnitude of a loan to deposit ratio for a savings association. Most vexing, however, is the question of whether statistical communities can reasonably be assumed to be target areas for Dallas County savings and loan associations. It is certainly possible that the instability of the loan to deposit ratios represents not only variations in institutional lending policy, but also an inappropriate definition of target areas.

With these limitations in mind, we examined the question of whether the variations in loan to deposit ratios form some sort of geographic pattern beyond the statistical community. The results once again lent support to the notion that loan to deposit ratios are highest in the far northern sections of the county. Indeed the imbalance seemed much greater than either the simple distribution of loans or the distribution of deposits. As noted above, one of the problems in the analysis of loan to deposit ratios is the possibility of very small savings association offices skewing the result. To see whether this was in fact the case, a second analysis was undertaken in which the computed mean of various loan to deposit ratios was multiplied by the value of loans purchased or originated for each office. The ordinal ranks provided by the index once again reinforced the view that loan to deposit ratios are in fact higher in the northern portions of the county.

MARKET VARIABLES AND LOAN ALLOCATION

While there is substantial evidence for a geographical pattern in the distribution of home credit, there is no evidence that this implies the practice of redlining or disinvestment per se. Even the most critical analysis of the lending institutions would not argue that loans ought to be distributed evenly over the city. Obviously, one would expect few home loans in industrial areas where there is simply very little housing. This type of neighborhood differentiation explains at least a portion of the observed geographic patterns noted above. For example, three industrial statistical communities combined have only twenty-three conventional mortgages for 1977. To compare these rates meaningfully, therefore, we need to consider whether these communities are in some sense equivalent on the sorts of criteria which define credit allocation.

When confronted with the differential allocation of home credit described above, representatives of lending institutions often claim that they represent the workings of the free market. In this section we examine the predictive power of two market-type variables. These are: median income and total number of households. We take the total number of households as a short-run limit to possible demand for home credit. That is, the greater the number of people who live in a given statistical community, the greater the likelihood of a demand for home credit. Median income is an attempt to capture a second critical market variable, i.e., whether a particular loan appears to be secure. We would expect that as the perceived security of the loan increases, so does the probability that the loan will be granted.

Table 3.4 reports the results of a series of multiple regressions which attempt to "explain" interstatistical community variation in home credit by our two market indicators. Not unexpectedly, the explanatory power of these variables is impressive. For conventional mortgages these two variables explain approximately 50% of the intercommunity variation in the total value of loans. They also explain over 50% of the intercommunity variation for home improvement loans. Only for FHA/VA loans does the predictive power of the market variables decline. From Table 3.4 we

TABLE 3.4   Market Predictors of Home Credit Allocation Within
Statistical Communities, 1977[a]

| | Market Indicators | | |
	Median Household Income, 1976	Total Number of Households, 1976	Adjusted R^2
FHA/VA Loans			
Number	.09	.40	.15
Value	.15	.45	.21
Conventional mortgages			
Number	.49	.52	.57
Value	.54	.38	.49
Home improvement loans			
Number	.47	.53	.57
Value	.53	.45	.54

a. Table reports standardized betas.

note that income and number of households explain only some 21% of the variation in the value of FHA/VA loans. This result is hardly surprising, however, if one remembers that FHA/VA loans are most common in tracts with a moderate amount of conventional credit. They were rare in tracts which had either a good deal of conventional credit or those with very little. This would suggest that there is perhaps a curvilinear relation between our market measures (particularly median income) and FHA/VA loans. Examination of the relevant scatterplots (not reproduced here) give some support to this view.

In addition to the explanatory power of median income and population, we were interested in whether the level of savings deposits held in a statistical community was able to predict the allocation of home credit. Thus, the regression analysis reported in Table 3.4 was repeated with a third independent variable, the total savings reported for savings and loan associations operating within each statistical community. The analysis revealed that deposits had no predictive power with respect to home credit. Indeed, the simple Pearson r between measures of home credit and deposits ranged from a high of -.06 (for number of conventional mortgages) to a low of -.02 (for number of FHA/VA mortgages).

Objection may be made to either market indicator. For example, the number of households may greatly overstate potential demand in areas which have predominantly rental housing units. However, it may also understate demand in areas which are experiencing new housing construction. A critical validity issue for the security measure is at what point loans achieve a satisfactory level of security. It is difficult to maintain that to insist on ever greater security is justifiable on purely market criteria. We would argue that at some point demand for increased loan security (particularly when measured by income) represents a potentially discriminatory rather than market variable.

Keeping these limitations in mind we have computed a series of predicted loan rates from the regression analysis described in Table 3.4. Predictions were compared with actual loan rates for each of the statistical communities, then aggregated to the county zone level. These residuals show clearly that market variables (at least those reported in this chapter) are not able to account for the wide variation in home credit across Dallas County. We found a now familiar pattern which shows that the statistical communities in the far northern section of the county receive substantial investment beyond what one would expect on the basis of median income and population. The aggregated residuals for the near-northern and far southern county zones suggest that, in general, those sections of the county receive approximately what we would predict on the basis of income and population. The "surplus" home credit which appears in the northern areas of the county seems in some sense to be drawn from the near-southern county zone.

POLICY IMPLICATIONS

In this chapter we have attempted to document several empirical features of the home credit market (as defined by savings and loan associations) for Dallas county. These include:

(1) Most savings and loan associations place primary emphasis on meeting the credit needs of the northern, particularly the far northern, sections of the county.

The relative emphasis on the northern areas of the county seemed to hold for almost all lending institutions, and was particularly clear for conventional credit, whether the credit took the form of mortgages or improvement loans. At the level of the county zone, even FHA/VA mortgages followed this geographic pattern, although, as noted in the text, there were some important variations in the geographical distribution of federally insured loans compared to conventional credit.

(2) There is little evidence that the amount of savings which have been deposited within the county predicts the amount of home credit targeted to specific geographic areas. This seems true at

the very broad county zone level, as well as the much more restricted statistical community level.

By definition this lack of any clear tie between deposits and loans indicates a process of disinvestment in those areas in which savings are not "returned" in the form of home credit. Of course, the entire analysis reported in the paper depends on the assumption that either county zones or statistical communities can adequately represent target areas for individual savings and loan associations. Also critical to the present discussion is whether disinvestment as measured by loan to deposit ratios should be taken as a problem of public policy.

(3) There is significant variation among Dallas County statistical communities in level of home credit which may not be the result of purely market factors.

This last generalization is certainly the most problematic of the three, given the crude nature of the indicators used in the analysis. Nevertheless, there does seem to be evidence that at least some of the "natural" development in the northern sections of the county is not due simply to the automatic working of a market process.

This perceived imbalance in private credit within metropolitan areas has led to a great variety of proposed governmental responses. Perhaps the most direct—and certainly the most controversial—is some form of direct credit allocation. This view is often put forward by representatives of community organizations who wish to force lending institutions to invest funds in their neighborhoods. As one might expect, representatives of lending institutions greet this prospect with something akin to horror. Beyond industry rhetoric however, serious questions have been raised about how government might face the enormous task of allocating home credit outside the market framework (Brunner, 1975). Those skeptical of direct credit allocation also point out that even much less drastic intervention into the housing market through programs like public housing, urban renewal and 235 mortgage subsidies have often seemed counterproductive. Whatever the vices and virtues of credit allocation, there seems little likelihood that such a policy will be politically viable in the very near future. Indeed there is hardly anyone at the national policy-making level who argues for such an approach. In all the national disclosure requirements which have been passed by Congress, there have been specific disclaimers that the regulations were to be interpreted as supporting other than "sound lending practices" in reviewing loan applications.

For the most part, federal policy with respect to redlining and disinvestment has been what one might term *market enhancement*. In particular, the notion involves increasing consumer information which is, of course, assumed to be perfect in most competitive market models. The *Home Mortgage Disclosure Act* allowed individuals and community organizations to obtain data on where individual institutions were making loans. *The Community Reinvestment Act* carries on this disclosure approach by requiring savings institutions not only to disclose where loans are given, but also to provide information on the target community as perceived by the institution. The purpose of these regulations was to spur the organization of consumer groups to withdraw support from savings institutions which were not supporting the community in which they were located. Indeed, there have been numerous attempts by community organizations to force savings institutions to increase their level of neighborhood investment on the basis of this disclosure data.[12] These pressures are likely to increase as *Community Reinvestment Act* data become more generally available.

In addition to simple disclosure, there have been several other interesting attempts to increase the flows of private credit into areas generally thought to be credit poor. A traditional approach which has been developed both at the federal and local level has been to modify the perception of risk through some sort of subsidy or insurance program. Of particular interest is the experience of Dallas with Neighborhood Housing Services. (NHS). Nationally, NHS organizations have attempted to increase the flow of home credit to specific target neighborhoods by increasing market information for both consumers and lenders. The NHS is a private corporation which attempts to match credit-worthy individuals with lending institutions which have indicated that they are, at least in principle, willing to make loans in the target neighborhood. Initially, the NHS began in one community in Dallas. Based on the widespread perception that the NHS had been successful in the original target community, two new area NHS organizations have recently been established. Some limited evidence that the NHS has had a positive impact may be obtained if the residuals from market predictions discussed above are broken down to the level of the statistical community. The community which contains the longest-running NHS program shows slightly more conventional mortgage credit than expected by the regression analysis reported in Table 3.4. The picture is, however, far from clear. The community reports slightly fewer FHA/VA mortgages and fewer home improvement loans than expected by the market indicators.

The numerous efforts of government at all levels to redirect private investment to credit-poor urban neighborhoods suggest an emerging political consensus that problems of redlining and disinvestment do pose threats to neighborhood vitality. For the most part, the analysis reported in this chapter may be used to support that view. What is not clear is the substantive policy conclusions that should be drawn from these findings. On the one hand, redlining may represent imperfections or error within the market system. If this is the case, programs like the NHS may provide an efficient and relatively low-cost approach to relieving the problems of disinvestment and redlining. This logic supports the language of the *Community Reinvestment Act* which suggests that lenders have a legal responsibility to some defined target community (or where there are branch offices, target communities), and that, at a minimum, an institution which is not active in that community needs to explain why it is not. While lack of demand or lack of secure loans would be an acceptable explanation, it would be so only after the lender shows that such loans could not be generated through some sort of affirmative lending program. It is important to recognize, however, that such policies may be inappropriate if the maldistribution of home credit results not from error in the market system but in the proper working of the market. If this view is valid, lenders may be correct in their fear that attacks on redlining are in fact challenges to the market system per se.

A NOTE OF CAUTION

Throughout this chapter there has been an implicit assumption that it would be desirable—if it were possible—to redirect private investment more evenly throughout Dallas County. Increased involvement of lending institutions within neighborhoods long thought to be credit poor is often associated with broader efforts at neighborhood redevelopment or revitalization. While there is much that is praise-worthy in these efforts, they are not without possible negative impacts on current residents of affected neighborhoods. Since urban revitalization is often associated with an increase in the proportion of middle-class residents, revitalization may actually result in displacement of low-income populations. If there are no efforts to aid those displaced, the result may be far more than simple inconvenience. For example, if the total stock of low-cost housing is significantly reduced, the overall price of such housing is likely to increase. Such problems of displacement might well be brought on by *successful* affirmative lending programs of neighborhood lending associations.

NOTES

1. This view was developed over the entire range of Park and Burgess's writing. For a discussion of the specific application of these biological terms to the study of

urban change, see Burgess, 1966.

2. For a discussion of the filtering process see Lowry, 1960; Freiden, 1968; Nourse, 1973; and Smith, 1970. For an interesting variation on the filtering hypothesis, see Harvey, 1973.

3. For overviews of the mortgage market, see Klaman, 1961; Starr, 1975; and Grigsby and Rosenburg, 1975.

4. This has been documented by a number of community groups which argue that this lack of credit represents an explicit policy on the part of home lenders. Examples of such studies include: Przybyiski, 1978; Texas ACORN (Association for Community Organizations for Reform Now), 1976; Center for New Corporate Priorities, 1975; and Devine, 1975.

5. This finding is general in the studies cited in note 4.

6. Once again a major source of data on this point is the empirical work undertaken by various community organizations. See McKee, 1975; Devine, 1975; New York Public Research Interest Group, 1975; and Northwest Community Housing Association, 1975. All of these studies have been published in Senate hearings on the *Home Mortgage Disclosure Act.* Similar studies have been cited by Bradford, 1979.

7. Congressional concern with the issue of disinvestment led to the passage of the *Community Reinvestment Act.* This act requires all deposit lending institutions to define a target credit community as well as disclose loan policies. For a more complete discussion of the *Community Reinvestment Act,* see National Training and Information Center, 1979.

8. For a review of the legal issues surrounding redlining and disinvestment, see Wisniews, 1977; Baptiste, 1976; Voyles, 1976; and Loyola Law Journal, 1975.

9. Only savings and loan associations seem to be under a legislative mandate to serve a particular geographic area. Nevertheless, it is clear that the role of mortgage bankers is a critical one, and that any attempt to understand the complete home credit market must include them.

10. Since Dallas-Fort Worth has a number of overlapping census tracts, there is a clear requirement in Regulation C that lending institutions indicate in which city a loan was made. Given the large number of savings and loan associations which operated in both cities, it was clearly of some importance that this be done. Unfortunately, only a few of the savings and loans complied with this regulatión in 1976, and none was willing to correct its statement. The Federal Home Loan Bank Board initially indicated that it knew of no provision in the law that required such city disclosure. Only after the author presented to the Home Loan Bank Board a copy of its own regulation did it agree that city disclosure was required. However, the board still refused to order individual savings and loans to correct their 1976 statements. Rather, lending institutions were informed that they should make disclosures by city in all future statements.

11. When the lending institution did not report data for the calendar year, data for the fiscal year which most closely approximated the calendar year were used. For a more complete breakdown of the empirical distribution of these data, see Hula, 1979.

12. Accounts of such attempts are reported in the National Training and Information Center publication *Disclosure.*

REFERENCES

BAPTISTE, K. E. (1976) "Attacking the urban redlining problem." Boston University Law Review 56 (November): 898-1019.

BENSTON, G. (1978) "The persistent myth of redlining." Fortune 97 (March): 66-69.

BRADFORD, C. (1979) "Financing home ownership: the federal role in neighborhood decline." Urban Affairs Quarterly 14 (March): 313-336.

––– and L. RUBINOWITZ (1975) "The urban-suburban investment-disinvestment process: consequences for older neighborhoods." Annals of the American Academy of Political and Social Science 422 (November): 77-86.

BRUNNER, K. (1975) Government Credit Allocation: Where Do We Go from Here? San Francisco: Institute for Contemporary Studies.

BURGESS, E. (1966) "The growth of the city," pp. 47-62 in R. Park and E. Burgess (eds.) The City. Chicago: Univ. of Chicago Press.

Center for New Corporate Priorities (1975) Where the Money Is: Mortgage Lending in Los Angeles County. Los Angeles: Author.

DEVINE, R. (1975) "Where lenders look first: a case study in mortgage disinvestment in Bronx County, 1960-1970," pp. 1267-1476 in Hearings before the Committee on Banking, Housing and Urban Affairs, United States Senate, 94th Cong., 1st Sess. on S 406.

FREIDEN, B. J. (1968) "Housing and national goals: the old policies and new realities," pp. 170-225 in J. Q. Wilson (ed.) The Metropolitan Enigma. Cambridge, MA: Harvard Univ. Press.

GRIGSBY, W. and L. ROSENBURG (1975) "The inner-city housing market," pp. 153-194 in W. Grigsby and L. Rosenburg (eds.) Urban Housing Policy. New York: APS Publications.

HARVEY, D. (1973) "Use value, exchange value and urban land-use," pp. 153-194 in D. Harvey (ed.) Social Justice and the City. Baltimore: Johns Hopkins Univ. Press.

––– (1977) "Government policies, financial institutions and neighborhood change in the United States," pp. 123-140 in M. Harloe (ed.) Captive Cities: Studies in the Political Economy of Cities and Regions. New York: John Wiley.

HULA, R. (1979) Housing, Lending Institutions and Public Policy. Dallas: Southwest Center for Economic Development, Univ. of Texas at Dallas.

KLAMAN, S. B. (1961) The Post-War Mortgage Market. Princeton, NJ: Princeton Univ. Press.

LONG, N. (1975) "The city as reservation." Public Interest 25 (Fall): 62-38.

LOWRY, L. (1960) "Filtering and housing standards." Land Economics 36: 362-370.

Loyola Law Journal (1975) "Note on redlining: the fight against discrimination in mortgage lending." 6: 71.

LYONS, A. (1975) "Conventional redlining in Chicago," pp. 342-377 in Urban-Suburban Investment Study Group, The Role of Mortgage Lending Practices in Older Urban Neighborhoods. Evanston, IL: Northwestern Center for Economic Affairs.

McKEE, D. (1975) "Supplementary study, housing analysis in Oakley, Bond Hill and Evanston: financial investment patterns," pp. 1479-1623 in Hearings before the Committee on Banking, Housing and Urban Affairs, United States Senate, 94th Cong., 1st Sess. on S 406.

MEGEE, M. (1968) "Statistical prediction of mortgage risk." Land Economics 65 (November): 461-469.

National Training and Information Center (1979) The Community Reinvestment Handbook. Chicago: Author.

New York Public Research Interest Group (1975) "Take the money and run," pp. 337-361 in Hearings before the Committee on Banking, Housing and Urban Affairs, United State Senate, 94th Cong. 1st Sess. on S 406.

Northwest Community Housing Association (1975) "Mortgage disinvestment in northwest Philadelphia," pp. 1199-1218 in Hearings before the Committee on Banking, Housing and Urban Affairs, United States Senate, 94th Cong., 1st Sess. on S 406.

NOURSE, H. O. (1973) "The filtering process," pp. 91-106 in H. O. Nourse, The Effect of Public Policy on Housing Markets. Toronto: Lexington Books.

PERRY, D. and A. WATKINS (1977) The Rise of the Sunbelt Cities. Beverly Hills: Sage Publications.

PRZYBYISKI, M. (1978) Perception of Risk—the Banker Myth: An Eight City Survey of Mortgage Disclosure Data. Chicago: National Training and Information Center.

SMITH, W. (1970) "Filtering and neighborhood change," pp. 64-89 in M. Stegman (ed.) Housing and Economics. Cambridge, MA: MIT Press.

STARR, R. (1975) Housing and the Money Markets. New York: Basic Books.

STERNLIEB, G. (1975) "The city as sandbox." Public Interest 25 (Fall): 14-21.

Texas ACORN (Association for Community Organizations for Reform Now) (1976) Investment and Disinvestment in Dallas Neighborhoods. Dallas: Author.

VOYLES, R. (1976) "Redlining and the home mortgage disclosure act of 1975." Emory Law Journal 25: 671-672.

WEINSTEIN, B. and R. FIRESTINE (1978) Regional Growth and Decline in the United States. New York: Praeger.

WISNIEWS, R. E. (1977) "Mortgage redlining: parameters for federal, state and local regulation." University of Detroit Urban Law Review 54: 367-429.

PART II

NATIONAL POLICY
Trials and Errors

4

Urban Renewal and the Revitalized City:
A Reconsideration of Recent History

HEYWOOD T. SANDERS

Urban renewal, the panacea for America's troubled cities in the 1950s and 1960s, has had its evident failures, bulldozing entire neighborhoods and replacing, in all too many cases, old but service-able houses with forbidding, multiple-unit monoliths that lacked humanity, scale, and any sense of community or architectural distinction.

In city after city, such projects rapidly became breeding grounds for crime and alienation, accelerating the decay of downtown districts and the flight to the suburbs of those who could afford to get out, aggravating the despair of those who could not.

—Barbara Diamonstein (1978:16)

Boston Mayor Kevin White slumps comfortably in the front seat of his official car and smiles with satisfaction at the visible progress his city has made. Just 100 yards from city hall is bustling Quincy Market—formerly a deserted waterfront area and now a crowded festival of chic shops and restaurants . . . "Boston," he says, "is incredibly healthy."

—Newsweek (January 19, 1979)

☐ FROM 1949 UNTIL 1974, the federal urban renewal program was a central element in national urban policy and public debate. Widely criticized as "Negro removal" and attacked for its destruction of urban neighborhoods and notable buildings, the program fell into disfavor both inside and outside of government. The current wisdom even suggests that renewal, with other federal programs of the time, may well have contributed to the decline and demise of the central city. In a period where renovation, rehabilitation, and reuse are enthusiastically supported, it is difficult to argue for a reexamination of a program originally designed to demolish and destroy. Yet urban renewal merits a review, for much of what is seen as signs of the revival of the city are its fruits, finally blossomed. The Quincy Market renovation in Boston, the Gallery shopping mall in downtown Philadelphia, the Charles Center and Inner Harbor projects in Baltimore all represent the results of urban renewal. The full economic and social meaning of these physical results is only now becoming apparent. Too late, we may have learned the mistakes of renewal so well that we are unable to duplicate its successes.

The reconsideration of the renewal program which follows will cover three general areas, from the vantage point of the middle 1970s. First, the design and overall impact of the program will be examined, including all efforts initiated through the end of the program in late 1974. The process of acquiring, clearing, and reselling land is a time-consuming one, and the contemporary image of renewal is likely to be a more substantial and accurate portrayal than earlier views. Second, changes in the physical character of renewal projects over time will be reviewed because changes in federal goals and regulations, particularly in the latter part of the 1960s, proved to have a dramatic impact on the character of the program. Third, an effort will be made to suggest the scope and form of local renewal efforts. Urban renewal was never a single, national program. Its implementation depended on the vagaries of local politics and pressures in over 1,000 separate entities of local government. Thus, the impact and physical shape of renewal showed considerable variety, both in terms of who it benefited and who it hurt.

PROGRAM DESIGN AND OUTCOMES

Enactment of the Housing Act of 1949 provided a strong symbol of national concern with the problems of slums and blight. However, the original program of slum clearance and redevelopment was not designed to aid the slum dweller directly but to alter the shape of urban land uses. This intent is demonstrated by the renewal program's two primary innovations—the use of eminent domain (through individual state enabling legislation) to acquire and clear land for private reuse; and a direct subsidy (the so-called write-down) on the cost of the land. No direct subsidies were provided for new construction or development. The program thus represented a curious mix of public (intergovernmental) and private responsibility. The federal government defined the outlines of the program, selected among the local projects submitted, and provided the bulk of the financing. Local governmental bodies selected the projects and their locations, directed the process of acquisition and clearance, and then made the land available to private developers. The burden of new development rested on the private real estate market, with profit as a central motivating force.

The inability of local governments to direct the development process proved to be a major limitation of the program. Localities could develop attractive new plans for development, but the timing and form of new uses rested outside their authority. Where markets for cleared land were strong, the renewal process moved quickly and effectively, although subject to the charge that unsubsidized private development might well have provided the same results. In communities (or individual project areas) where little or no market existed, the program's legacy was often acres of barren lots.

Local governments suffered from two central weaknesses. First, they lacked a great deal of information, particularly about the market for land on which the ultimate results of their efforts rested. The alternative of doing only what was acceptable to local developers often meant modest projects of limited impact and distinction. The federal government attempted to correct this particular information deficit by encouraging a variety of economic and market studies prior to the initiation of most projects.[1]

At the same time, however, local governments suffered no lack of demands from powerful constituencies. The hotel owner who found skid-row encroaching on his investment or the hospital or university that demanded room for expansion, parking, and residences with the threat of a move to the suburbs sought to gain a renewal project that would benefit their immediate interests. In many cases, the central interest was in getting rid of some source of aesthetic or social unpleasantness with little regard for its replacement. Local politicians also needed to demonstrate their leadership with announcements of imaginative and dramatic projects at regular intervals, even when they had little interest in their eventual shape or success. The federal administrators of urban renewal simultaneously sought to enlarge both their domain and their popular support by promoting and accepting such short-term-oriented renewal efforts. In the final analysis, the implementation of urban renewal programs rested on a trust in local agencies to make wise decisions about such matters as land uses or the form of the city in twenty years, about which they knew and cared very little.

Such things as delay, vacant and undeveloped land, and the displacement of the poor are generally seen as programmatic failures. They may be the natural and inevitable result of the operation of these local political processes. Similarly, displacement's impact on the poor is not a necessary function of the design of the program. Rather, it is a result of an inequitable distribution of political power at the local level, uncontrolled and unaltered by the federal government.

Yet, criticism of the urban renewal program, from both the left and the right, has been sufficiently widespread to present a solid impression of misguided national activity. Frieden and Kaplan (1975: 22-25) summarize many of the points repeated by the critics. First, a program which was reputed to be a vehicle for improving housing had, in fact, become a means for cities to "refurbish the central business district, build housing for middle- and upper-income families, and bolster their property tax base." Second, the program's impact on housing posed a particular burden for lower-income families. As Scott Greer (1965: 3) noted, "At a cost of more than three billion dollars, the Urban Renewal Agency . . . has succeeded in materially reducing the supply of low cost housing in American cities."

While each of these characterizations was not an entirely inaccurate description of urban renewal in the early 1960s, they provide a false impression of the program at the time of its demise. Local pressure to increase the property tax base and attract valuable new land uses often supported a shift away from housing, particularly for low-income families. However, as of December 1974 (the last date for which information is available), new housing represented the single largest new use of cleared renewal land. Over 35% of the 98,000 acres acquired under the program was designated for new, privately owned dwellings. A further 5.2% was devoted to new public housing. In comparison, only 28.4% of the redeveloped acreage was planned or actually used for new commercial and industrial structures. Were this new housing development largely designed for upper-income persons, criticism would still be warranted. In fact, of the new housing units actually underway or completed from 1950 through June 1972, almost 55% were specifically designed for low- and moderate-income families, using both federal and state subsidy programs.

The failure of the renewal program to provide new housing has been argued with regularity. Martin Anderson notes in the *Federal Bulldozer:*

> In essence, the federal urban renewal program eliminated 126,000 low-rent homes, of which 80 percent were considered to be substandard, and replaced them with about 28,000 homes, most of them in a much higher rent bracket. Thus, the net effect of the program has been to aggravate the housing problem for low-income groups and to alleviate it for high-income groups [1964: 67].

While his figures characterize the program as of the middle of 1961, the relative gap between demolitions and new construction has been a consistent argument against the program. Such an outcome is inevitable in a program that demolishes housing as its earliest activity, and as a precondition for any new development. Problems of marketing cleared land compounded this natural delay, with the result that some four years, on average, elapsed between the beginning of land acquisition and the first new construction activities. This situation was worsened by the growth of the program, for new construction on a modest number of projects was being compared to the volume of demolitions on a much larger group. For example, of the 578 projects which were underway by the end of 1961, half had been initiated since 1959.

A more reasonable picture of the impact of renewal on the housing stock can be gained by comparing demolitions with the total housing production to date. The roughly 570 projects underway in 1961 included some 224,000 families and a slightly larger number of housing units. By 1974, this group of projects contained some 182,600 new dwelling units, including almost 16,000 new public housing units. While renewal clearly accounted for a net decrease in housing, that decrease was far less than often suggested.

The impact of renewal on families and individuals, through relocation and displacement, is perhaps the single most damned element of the program. There can be no doubt that displacement often carried significant economic and psychological burdens. Yet, the image of "Negro removal" is not a completely accurate one. From 1950 through mid-1971, 293,068 families and 157,297 single individuals were relocated by local renewal authorities. Of those families for whom race was recorded (91% of the total), blacks and nonwhites constituted 57.9%, whites roughly 37%, and other minorities (largely Hispanics) 4.8%. For the single individuals affected, the racial burden was shifted. The majority of relocated persons were white (55.5%), with blacks totaling 41%. Thus, while blacks bore a disproportionate burden of the costs of displacement, they were not its exclusive recipients.[2]

The impact of urban renewal on minorities should also be assessed in light of some further data. The racial balance of relocation was gradually altered as the program itself changed during the 1960s. As a result, only 46% of the persons displaced during 1973 and 1974 were black, with Hispanics representing another 13%. The racial figures may also conceal substantial variation in local programs, as cities differed in both racial concerns and political responsiveness. Lastly, just as the costs of renewal were distributed inequitably, so was one of the major benefits. Of the 250,000 new housing units on renewal sites for which occupancy data were available, blacks occupied over 101,000 units, or almost 41%. Other minority groups accounted for an additional 5.8% of the occupied units.

A view of urban renewal from the mid-1970s clearly suggests some additions and amendments to the widely quoted reports of the previous decade. With an average of some ten years between the planning and completion of a typical project, *time* is a necessary element in a proper evaluation of the program. The inability of urban renewal to deliver its new, physical results quickly may have been its most important failure, enhancing the arguments of its opponents. At the same time, these critical analyses undoubtedly assisted the process of change that altered the program during the last half of the 1960s.

PROGRAM CHANGE

During much of its history, the federal government was never able to clearly specify the goals of urban renewal. It began as a program of "slum clearance and redevelopment," with support from those concerned with the problems of housing the poor as well as those who sought a bold tool for renewing cities. A "predominantly residential" requirement effectively limited projects to either slum clearance or the development of new housing sites. By 1954, a shift of name and strategy to "urban renewal" provided for the options of rehabilitating residential areas or undertaking purely commercial redevelopment efforts (under an exception to the residential requirement). From 1954 until the mid-1960s, successive legislative amendments broadened the application of renewal and increased the number and size of exceptions to the residential emphasis.

By 1966 the climate surrounding the renewal program had substantially changed. Efforts to expand the size of the nonresidential exception were blocked, and the Congress for the first time required housing-oriented projects to specifically provide for low- and moderate-income families. The most important shift came in 1967 when the Secretary of the Department of Housing and Urban Development (HUD), Robert Weaver, announced three national goals for the program:

(1) the conservation and expansion of the housing supply for low- and moderate-income families;
(2) the development of centers of employment for the jobless, the underemployed and lower-income persons; and
(3) projects serving areas of "physical decay, high tensions, and great social need" [U.S. Congressional Research Service, 1973: 46].

Secretary Weaver's statement of goals gave the program a clear social focus for the first time since 1954, and influenced the program's shape until its demise in 1974.

The impact of changing federal strictures on renewal should be visible in terms of the characteristics of land use in renewal projects. The available national data are taken from HUD's unpublished reports which divide renewal project land into five categories of use (1) public residential, (2) private residential, (3) public institutional, (4) private commercial and industrial, and (5) streets and rights-of-way. The last category varies little by project, and will not be examined here.

Each individual project can be characterized in terms of its land uses *before* acquisition and clearance, and in terms of new uses, both planned and actual. The extent of slum clearance is indicated by the proportion of acquired land which was in private residential use before renewal. The production of new housing is similarly indicated by the percentage of cleared land redeveloped for private residential purposes. In both of these cases, the base is the total number of acres cleared and redeveloped in each project.

The shift from "bulldozer" clearance to rehabilitation and spot clearance is indexed by a different measure, the percentage of residential land prior to renewal which is designated for rehabilitation. The base for this measure is the total of all residential land in a given project, whether designated for clearance, rehabilitation, or no action at all. The rehabilitation percentage is thus calculated independently of the acquisition and clearance measures, and cannot be directly compared with them. Although some cities employed rehabilitation in commercial areas, only housing rehabilitation will be considered here.

These measures can be illustrated by some individual projects. The West End project in Boston, described by Gans in *The Urban Villagers* (1961), was largely residential prior to renewal. Of its 47 acres, over 20 (44%) were occupied by private housing. This 20-acre area was totally cleared; no rehabilitation was attempted. Despite the vast impact of clearance, the project provides roughly the same portion of its area to *new* housing (44%) as before renewal. However, almost all of the roughly 2,400 new dwellings in the West End are designed for middle- to upper-income occupancy, with only a 150-unit elderly project directly aided by federal subsidy. Another Boston project, the Downtown Waterfront area, represents a marked contrast to the West End. Its 60 acres of clearance did not include any residential area. Thus, its value on the residential acquisition measure is zero. Some of this cleared land—17 acres, or almost 29%—however, is devoted to new housing, largely luxury apartments. Finally, Boston's Charlestown project provides an example of a rehabilitation effort. Of almost 130 acres of residential land prior to renewal, only 24% was designated for clearance. Thus, the Charlestown project would "score" 76 percent on the rehabilitation measure. The project also included the development of over 800 new housing units as of 1972, with over 90% either public housing or subsidized private units for lower-income families.

The impact of changing federal strictures on renewal should be visible in terms of characteristics of land use in renewal projects nationally. The average percentages over time are presented in Figure 4.1.[3] The three measures of renewal land use (housing clearance, new housing construction, and housing rehabilitation) have been averaged by the year in which project planning was begun. Each urban renewal project is weighted equally, regardless of size or cost.

A heavy emphasis on slum clearance characterized the program between 1950 and 1956. From that point until 1967 there was a steady and regular decline in residential clearance, with a low point in the middle 1960s. The impact of Secretary Weaver's decision, and later housing legislation, is both immediate and dramatic. The concentration on residential neighborhoods increased by the early 1970s to a point roughly equal to figures for the early 1950s. Clearly, the statement of national renewal goals was reflected in a turnaround of local programs and activities.

The shift in residential emphasis was partially paralleled by a change in the way residential neighborhoods were affected by the renewal process. The availability of a rehabilitation alternative following passage of the 1954 Housing Act prompted a limited shift away from clearance. From 1956 through 1965, the average project had roughly 10% of its residential land area devoted to rehabilitation. In reality, the difference in size between a rehabilitation project and a clearance one meant that a far larger proportion of total land and homes was upgraded than was demolished. For example, projects begun in 1964 included 2,922 acres of housing to be cleared and 1,540 acres of housing rehabilitation, or a ratio of 1.88. By 1972 the relationship had shifted substantially, with 2,875 acres of clearance and 10,386 acres of rehab, a ratio of .28. An increase in rehab activity in 1966 presaged the shift in national goals, and marked the beginning of a substantial increase in rehabilitation. This figure reached 30% by 1971, some three times greater than the figures for the 1955-1965 decade.

A third index of renewal activity is shown in Figure 4.1 for the percentage of cleared land devoted to new housing. Sharp swings in housing development in the early years of the program reflected both the limited number of project starts and the impact of the "predominantly residential" requirement. With a high level of slum *clearance*, cities were free to build nonresidential structures on renewal sites. From 1958 to 1965, the program shows a more regular pattern of decreasing residential construction as the nonresidential exceptions were steadily enlarged. A sharp upswing in housing comes in 1966, with a dramatic and continuing rise to the early 1970s. Despite drops in 1971 and 1973, the percentage of land devoted to new housing development in the 1970s was greater than at any other time in the history of the urban renewal program.

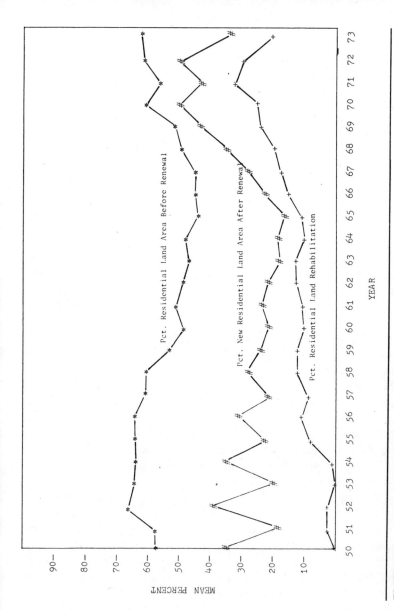

FIGURE 4.1 Urban Renewal Project Characteristics by Year

111

The shift to housing concerns in 1966 also saw an increase in the income groups served by new dwellings. During the 1950s, only about 30 to 40% of new residential units were designed for low- and moderate-income occupancy. Most of those lower-income units were developed in New York City under a variety of state and local subsidy programs. By the 1967 to 1970 period, the majority of all new units (between 60 and 70%) were developed under federal subsidy programs aiding families with below-average incomes. Unfortunately, the imposition of a moratorium on new subsidized housing commitments in January 1973 effectively limited the promise of new low-income housing on cleared renewal sites. As one result, many communities which responded to the change in national goals found themselves with all but unmarketable land which had been designed for new housing that could not now be built.

The achievement of national goals could be judged only a limited success if their imposition was resisted or deferred by the localities involved. The rate at which new projects were started provides a useful index of local response to urban renewal opportunities. After a slow start in the 1950s, an average of 158 projects were initiated each year from 1960 through 1966. The figures for the later years of the program suggest a positive local response to the federal concerns. While 1967 saw fewer new project starts (138) than the peak years of the decade, activities increased to a new high in 1970, with 197 projects begun. A sharp cutback in new federal commitments in 1971 was followed by the largest number of new project starts in the program's history—277. Clearly, even in the face of a new federal "social emphasis" that reduced the freedom of local renewal agencies, cities wanted and needed urban renewal.

The combination of an increase in new projects and a shift in their character meant that the physical shape of urban renewal was totally new. From 1968 through 1974, new renewal projects added over 14,750 acres solely for the development of new housing, or roughly 43% of the total housing acreage made available during the program's 25-year history. Another 31,000 acres of land contained residential properties to be rehabilitated. This vast area encompassed over 63% of all the housing rehabilitation area in the history of urban renewal.

The data in Figure 4.1 suggest that urban renewal efforts should be divided into two broad "eras." The first, from the beginning of the program through 1967, was marked by an ever-increasing freedom for local choice in project selection with little federal dictation. From 1968 through 1974, the clear statement of national goals reduced local discretion and shifted the choice process toward federal objectives. The division of renewal activities into these two distinct periods provides some analytical benefits. First, individual cities can be examined for their responsiveness to the new national goals. Second, any analysis of local choice and decision-making can focus on pre-1968 period when *local* rather than national policies determined the form of city renewal actions.

Table 4.1 illustrates the change in local programs for a selected group of large cities. In general, they show a high level of response to national program changes. Philadelphia provides a case in point. As shown in Table 4.1, that city responded dramatically to the shift in program concerns. During the first renewal era, through 1967, the city showed moderate levels of both housing clearance (41%) and residential rehabilitation (22%). The period from 1968 through 1974 saw an increase in residential emphasis, together with a dramatic upswing in the level of rehabilitation to an average of 68%. The shift in federal regulations also supported a sharp change in the pattern of land uses. In terms of new land uses, Philadelphia witnessed a three-fold increase in new residential land and a corresponding *decrease* in the average proportion of projects and land devoted to commercial reuse. A similar pattern is evident in such places as Minneapolis, Baltimore, and Portland. In some cases, new program priorities are evident in only selected areas. For example, Chicago's program saw an increase in residential land from 30 to 75%, although both housing acquisition and rehabilitation declined. In New York City and Boston, the federal changes

TABLE 4.1   Residental Land Use Before and After Renewal and Housing Rehabilitation for Selected Cities

	Average Percentage of Land Residential Before Renewal		Average Percentage of Residential Land for Rehabilitation		Average Percentage of Land Residential After Renewal	
	Before 1968	After 1968	Before 1968	After 1968	Before 1968	After 1968
Atlanta	67	61	19	9	24	29
Boston	31	24	41	52	32	22
New York City	44	51	6	0	62	54
Philadelphia	41	52	22	68	22	64
Baltimore	48	51	15	24	20	50
District of Columbia	39	35	12	32	43	60
Kansas City	63	59	21	26	32	74
Minneapolis	69	75	34	50	48	83
Denver	42	60	33	18	26	51
San Francisco	33	77	8	0	38	59
Portland	52	74	30	35	24	51
Chicago	54	39	8	1	30	75

Note: Classification based on year project planning began.

appear to have had only limited impact. Local political pressures and needs, together with a previous history of new housing development in the New York case, moved the program in directions other than those preferred by Washington.

PROGRAM IMPLEMENTATION AND LOCAL OUTCOMES

Between 1950 and 1967, the available scope of local choice was quite broad. Thus, the physical shape and impact of urban renewal could vary substantially from one community to the next. The balance of this discussion will deal with the nature and causes of variations among localities. In order to eliminate the impact of federal-level decisions, only those projects initiated before 1968 will be considered.

The data on large cities displayed in Table 4.1 indicate the variety of local renewal undertakings. In terms of slum clearance, Atlanta and Minneapolis stand out clearly, with the average project largely devoted to residential land acquisition. They present a sharp contrast to Boston and San Francisco, where residential land was only one-third of the cleared area of a typical project before 1968. Even these similarities conceal a range of differences in approach and treatment. Minneapolis shows almost twice the level of residential rehabilitation coverage of Atlanta prior to 1968. In terms of total land area, Minneapolis upgraded almost one acre of homes for each acre it cleared. Atlanta's ratio stood at two cleared acres for each rehabilitated. Thus, Atlanta's activities show a clear tendency toward "bulldozer" clearance of residential neighborhoods. The Minneapolis figures suggest that its residential emphasis included a mixture of residential rehabilitation and demolition.

In contrast, in Boston and San Francisco, the emphasis was on nonresidential areas. In both of these cities urban renewal was heavily used in and near the downtown commercial core. However, Boston mixed its downtown program with a massive program of neighborhood housing rehabilitation, in contrast to the efforts at housing clearance in San Francisco's Western Addition and Hunters Point areas.

Cities can also be differentiated by the level of new housing development on renewal sites. New York City clearly stands out. With 62% of land devoted to private housing construction, including some 48,000 units as of 1972, New York City accounted for almost 15% of all the new dwellings built on renewal land. Atlanta, Philadelphia, and Baltimore show a far more modest interest in new housing, averaging under a quarter of project land. For some cities, such as Minneapolis, new housing provided the major reuse of land cleared as part of a larger neighborhood rehabilitation effort.

There are also notable differences in the beneficiaries of new housing production. Some cities pursued the construction of low- and moderate-income housing even in the absence of federal constraints. On the average, 61% of the new private housing development in Boston received some federal subsidy targeted to the lower ranges of the income spectrum. This figure reached 85% in Denver and 100% in Kansas City. In Portland, Philadelphia, and the District of Columbia, housing was aimed at higher-income families. An average of only 36% of new homes in Portland was subsidized; the figures for Philadelphia and the nation's capital were 42 and 40%, respectively.

Urban renewal efforts can also be classified by the new, nonresidential uses of cleared land. Public nonresidential uses include such things as new schools, city buildings, and civic centers. Private nonresidential developments run the gamut from industrial structures to offices and shopping facilities. The Government Center project in Boston's downtown core involved roughly 44 acres of clearance. With a new city hall, and state and federal office buildings, new public nonresidential uses total 62.5% of the project area. Privately owned developments account for a further 20.3%. In contrast, the city's Downtown Waterfront project of over 60 acres is dominated by private commercial reuses.

Table 4.2 displays both new public and private nonresidential land uses in a number of localities. New, tax-exempt public developments are notably *absent* in Minneapolis and New York City. Portland, Baltimore,

TABLE 4.2   Public and Private Nonresidential Land Use After Renewal for Selected Cities

	Average Percentage Public Nonresidential After Renewal		Average Percentage Private Nonresidential After Renewal	
	Before 1968	After 1968	Before 1968	After 1968
Atlanta	44	24	20	22
Boston	17	9	34	20
New York City	9	27	15	12
Philadelphia	15	17	47	19
Baltimore	46	15	24	20
District of Columbia	14	7	32	32
Kansas City	21	1	39	14
Minneapolis	7	2	30	7
Denver	22	23	42	0
San Francisco	11	0	36	39
Portland	48	25	14	17
Chicago	24	6	39	15

Note: Classification based on year project planning began.

and Atlanta are clearly dominated by these new public uses. These cities have proved willing to spend large sums on new civic buildings despite the fact that they return little in terms of property taxes. Public reuse is, in this sense, perhaps the most expensive new use of renewal land.

Privately owned nonresidential uses constitute a common and more inclusive category of renewal land use, ranging from new industrial buildings to offices and hotels in the central core. Most of the large cities noted in Table 4.2 show substantial new taxpaying nonresidential development. New York, with a strong emphasis on housing, and Portland, with such public redevelopment as a city auditorium and state college, both show low private nonresidential activity. Philadelphia and Denver demonstrate a more consistent interest in private renewal, with large new commercial and office developments in and near their downtowns and sites for private university expansion.

The data in Tables 4.1 and 4.2 indicate something of the range of variation in local renewal programs. From a concentration on new housing to an emphasis on office development, from a sensitive use of housing rehabilitation to total clearance and demolition, cities have been able to tailor this federal program to their own needs and political environments. The wide variety of city types, on such dimensions as age, governmental structure, race, and growth, allows for major differences in the perception of civic problems and the ability to act on them.

A previous analysis of local renewal programs indicates that regional location is the most important correlate of the form of renewal (Sanders, 1977). Geographic region taps a number of dimensions that affect clearance and housing policy. The age of housing and commercial facilities, the presence of a large minority population, and the differences between machine and reform styles of local government—all vary by region and exert an impact on local renewal strategies.

Tables 4.3 and 4.4 present the regional means for the same indices of land use and new development discussed earlier.[4] These means are unweighted, with each project counted equally. They are therefore not dominated by the larger and more expensive projects in major urban centers. The regional classification follows the demarcation used by the Inter-University Consortium for Political and Social Research, with a clear distinction between states in the South and border areas.[5]

The examples provided by Boston, Minneapolis, Atlanta, and Portland appear to characterize the renewal actions of their respective regions. The South, much like the city of Atlanta, stands out with a high level of clearance activity in residential neighborhoods. The Southern slum clearance emphasis is intensified by a low level of residential rehabilitation. Projects in the Border states also show a strong concern with residential

TABLE 4.3   Residential Land Use Before and After Renewal by Region[a]

	Average Percentage of Land Residential Before Renewal	Average Percentage of Land Residential Rehabilitated	Average Percentage of Residential Land After Renewal
New England	41.2 (179)[b]	12.8	17.7
Middle Atlantic	43.3 (529)	9.2	24.1
E. North Central	52.6 (281)	13.0	26.0
W. North Central	49.0 (118)	20.1	22.1
Southern	59.2 (352)	9.6	21.2
Border	53.7 (167)	9.1	16.5
Mountain	50.4 (18)	12.3	23.1
Pacific	43.9 (97)	7.2	25.1

a. See Note 5 for regional breakdown.
b. Number of projects are indicated in parentheses.

areas unmitigated by rehabilitation. In many other regions, the impacts of renewal are more strongly felt in commercial zones. New England and the Middle Atlantic and Pacific areas show relatively low levels of housing acquisition. The alternative of commercial and business-oriented renewal affects both the politics and results of renewal activity. Potential resident opposition can be avoided with a concentration on commercial areas, and such areas provide the best sites for high-value commercial and residential development.

Residential rehabilitation reaches a peak in the West North Central region, and particularly in Kansas and Minnesota. Rehabilitation can provide a more politically palatable mode of intervention in residential neighborhoods than total clearance, and it may be the only renewal mechanism possible in cities where neighborhood interests are strong or local government weak.

Land for new privately owned housing is a common product in almost every geographic region. Although the Border and New England states show slightly lower levels of new residential land, there is a remarkable consistency across the country. This may reflect the constraints of the early period of slum clearance and the "predominantly residential" requirement. It may also reflect the tremendous variety of products within

TABLE 4.4   Public and Private Nonresidential Land Use After
Renewal by Region[a]

	Average Percentage Public Nonresidential After Renewal	Average Percentage Private Nonresidential After Renewal
New England	20.8	42.9
Middle Atlantic	22.6	37.5
E. North Central	20.7	39.1
W. North Central	20.8	43.2
Southern	25.1	27.6
Border	31.2	31.0
Mountain	31.2	31.0
Pacific	17.6	41.4

a. See Note 5 for regional breakdown.

the realm of new housing. Some locales emphasized new middle- and upper-income developments to lure surburbanites back to the central core. Other areas developed dwellings for lower-income families, either as a result of clearance efforts or as part of broad-scope rehabilitation and neighborhood conservation programs. New housing development may result from local political forces that seek to alter the racial or economic character of a project area without changing its residential nature.

In the area of private nonresidential uses, as shown in Table 4.4, New England ranks relatively high. This reflects a concentration on the commercial redevelopment of downtown areas in many cities in Connecticut (averaging 44% new private nonresidential uses) and Massachusetts (averaging 46%). Nonresidential renewal in the West North Central region includes substantial industrial redevelopment, as well as other commercial new uses. New private nonresidential development is also substantial in the Pacific area. The similar character of urban renewal in the East and Far West is notable, for cities in these areas are quite different in terms of age, development patterns, and economic history. Private nonresidential development is particularly limited in the South. The lack of new commercial and industrial construction in this program probably reflects both a strong emphasis on residential slum clearance and an unwillingness to use the powers of eminent domain in commercial and downtown areas.

Public nonresidential development presents something of a mirror image to private activities. The Far West and East show low levels of nontaxpaying institutional construction, in comparison with the Border and Mountain regions. The Border region average is largely dominated by a single state and its largest city. The public reuse figure for Maryland is

40%, with the city of Baltimore at 46%. These values reflect an early emphasis on slum clearance and support for institutional expansion at the University of Maryland and Johns Hopkins Hospital.

The high level of public redevelopment in the Mountain area should be interpreted with caution, due to the small number of renewal projects in the region. One mountain city, Tucson, Arizona, does illustrate some important aspects of public nonresidential use. Urban renewal planning began in Tucson in 1958, with plans for a 400-acre project. Political opposition of all sorts effectively stymied local action, and it was not until 1967 that execution began on the city's first project. The final form of Tucson's renewal was a 76-acre project devoted almost entirely to new public uses, including a civic center and convention hall, and new county and city office buildings. Tucson was successful at this form of renewal largely because new public buildings aided major downtown property owners without threatening their commercial interests or providing any competition to store owners (Joyner, 1967: 185-192).

New civic developments are often popular with local business interests. They provide a center of public activity in the downtown core without creating competititive office or retail space. In Atlanta (with 44% public nonresidential reuse), Birmingham (46% public), and Norfolk (45%), the new civic center/convention hall proved a useful replacement for close-in slum neighborhoods and decaying homes which could garner the support of both politicians and businessmen.

LOCAL PROGRAM CHOICE AND THE
POLITICS OF DISPLACEMENT

An analysis of displacement and relocation is complicated by the shifts in federal goals and reduced local discretion after 1967. In addition, the passage of the Uniform Relocation and Real Property Acquisition Act of 1970 altered both the benefits received by displacees and the available records on relocation. The following discussion will therefore examine relocation initiated by June 1971, which eliminates many of the renewal projects begun after 1968-1969.

From 1950 through mid-1971, 293,068 families and 157,297 single individuals were relocated by local renewal authorities. An additional 102,012 businesses were forced to move as their locations were acquired. Thus the impact of the program did not fall solely on families. The shifts toward commercial clearance and redevelopment in the early 1960s simultaneously shifted the character of displacement to affect businesses located in or near the central business district.

The balance of this discussion will examine the character of local relocation using four measures. The percentage of business displacement indicates the relative proportion of total business and family relocation made up by firms. A second measure, the percentage of individual displacement shows the relocation impact on single individuals relative to the sum of family, business, and individual displacement. Two additional measures suggest the precise impact of renewal. An index of racial selectivity is provided by the proportion of family relocation involving blacks and other nonwhites. For business firms, a measure of the percentage of displaced firms that went out of business indicates both the economic health of displaced firms and the character of the renewal impact.[6]

The variety of local renewal efforts is indicated in Table 4.5, which includes data for the sample of large cities previously discussed. The first measure, the business displacement percentage, varies from 11.5% in Atlanta to almost 45% in Denver. The value for Atlanta clearly reflects that city's emphasis on residential clearance. This family displacement and slum clearance was directed almost entirely at black families, for close to 90% of all relocated families were nonwhite. The almost total emphasis on family relocation is notable in Atlanta, but Chicago, New York, and the District of Columbia also demonstrate a strong residential clearance and family displacement bias. For these three cities, high rates of family displacement also included an overwhelming emphasis on minorities. The concurrent impact of high family displacement and great impact on minorities in the nation's largest cities is perhaps the central factor in the

TABLE 4.5   Relocation Characteristics for Selected Cities (percentages)

	Business Displacement	Individual Displacement	Nonwhite (families)	Businesses Discontinuing Operations
Atlanta	11.5	23.6	88.6	34.8
Boston	35.6	27.0	47.9	27.2
New York City	18.3	20.9	a	25.8
Philadelphia	22.8	33.8	79.9	16.2
Baltimore	21.6	22.2	83.0	22.7
District of Columbia	17.4	21.2	85.6	45.7
Kansas City	25.3	38.0	56.5	33.9
Minneapolis	30.1	47.5	36.3	39.6
Denver	44.7	32.9	43.8	34.5
San Francisco	34.9	45.2	79.5	39.0
Portland	39.0	63.1	23.6	46.0
Chicago	17.1	33.7	78.7	31.6

a. The value for New York is not shown due to the large number of families for whom race was not reported.

image of urban renewal as Negro removal. Yet this combination is not necessarily the only form of a local program. Boston, San Francisco, Portland, and Denver suggest that businesses may bear a fairly large share of the burdens of acquisition and displacement. For these four cities, the political and social problem of family displacement (particularly involving minorities) may have supported an alternative approach to renewal which emphasized commercial areas rather than residential neighborhoods.

The renewal of business areas provides all the benefits of increased taxes, new construction, and bold and visible public action without disturbing a mass of families and voters. Yet those cities which elected a commercial emphasis show additional dimensions of variation. Boston and Denver show low levels of both family displacement *and* individual displacement. In Portland, San Francisco, and Minneapolis, individuals bore much of the brunt of the displacement. The project histories of these cities suggest that this high individual displacement was the direct result of a focus on the local "skid row," and its single male occupants. This skid row emphasis is also mirrored in the figures on business failures due to displacement. In Minneapolis and Portland, the high rate of failing business firms reflects the modest economic health of those enterprises dependent on the residents of a skid row. The figures for Portland are largely the result of the South Auditorium project, which renewed a skid row neighborhood while displacing close to 300 businesses and almost 900 individuals. The lower proportion of firms out of business in San Francisco is a function of the large fraction of businesses in the Yerba Buena project area which were not affected by mid-1971. Inclusion of the most recent business displacement figures for Yerba Buena, where 45% of the 534 firms failed, would raise the business discontinuance proportion for San Francisco to approximately 42%.[7]

If Minneapolis and Portland tended to concentrate their efforts solely on the local skid row, programs in Boston and Denver had a larger impact on the downtown core area. The levels of individual displacement and business failures in Denver are a reflection of the city's Skyline project, located in the western portion of the central business district. The Skyline area included a mixture of substantial downtown firms and weaker enterprises, with a somewhat smaller population of single individuals than San Francisco or Portland. Boston's efforts involved a more substantial group of business firms in the downtown area. Only 25% of the businesses displaced in Boston's Downtown Waterfront project failed, and the rate for the city's Government Center project, which included 859 firms, is an even lower 17%. While skid row is generally considered unpleasant and its residents lack political power, the downtown core is a repository of far more economic and political power. Intervention in the core area is undoubtedly more difficult politically, but it is likely to provide far greater long-term benefits, both in terms of the city's property tax base and the value of adjacent downtown real estate. This form of renewal can only take place when a powerful political interest, either the dominant business firms (as in Denver) or elected leaders (as in Boston), are willing to invest both time and substantial political capital. The apparent similarities between Boston and Denver, however, should not be overstated. Boston's renewal efforts resulted in the displacement of over 4,100 businesses, while the Denver program caused the dislocation of only 480.

The sample cities suggest both the scope of displacement variation and some of its correlates. The analysis of regional patterns can help extend these results to the entire nation. Table 4.6 clearly demonstrates the relationship between geographic region and renewal displacement. Of the eight regions, the South shows by far the lowest level business displacement, and the greatest concentration on moving families. Business impacts reach high levels in the New England, Mountain, and Pacific areas. Together with its concentration on families, the South shows the most substantial rate of black and other nonwhite relocation, roughly 81%. Indeed, 27% of all nonwhite relocation occurred in the ten states of the South. The problems in drawing renewal project boundaries and the complex nature of Southern housing patterns might have been expected to insure that some white families would be affected even by programs aimed at the black residential area. Yet the relocation figures for many Southern cities show a unique impact on the minority population. For example, 95% of Birmingham's renewal program involved nonwhite families, with an equivalent figure in Savannah. In Richmond and Roanoke the percentage of nonwhite relocation reached 99%. Columbus, Georgia, and Charlotte, North Carolina, managed perfect records. They managed to relocate

TABLE 4.6   Relocation Characteristics by Region[a] (percentages)

	Business Displacement	Individual Displacement	Nonwhite (families)	Businesses Discontinuing Operations
New England	37.6	27.3	30.8	25.6
Middle Atlantic	31.7	27.6	46.8	27.0
E. North Central	22.3	20.1	66.0	31.2
W. North Central	28.6	33.5	45.1	29.7
South	16.9	23.1	80.9	28.2
Border	24.9	25.2	68.2	28.4
Mountain	39.8	33.4	27.9	35.9
Pacific	38.2	45.2	48.6	39.7

a. See Note 5 for regional breakdown.

roughly 520 and 1,110 families, respectively, without disturbing a single white family.

A sharp contrast to the South is provided by New England and the Mountain and Pacific regions. With a higher level of business displacement, families bore a relatively smaller portion of costs of renewal in these areas. New England and the Mountain states are also notable for their low level of nonwhite displacement. However, an additional 22.8% of the relocated families in the Mountain area were classified as other minority—largely Hispanics.[8] These racial figures again suggest a substantial variation in "Negro removal" by region, reflecting not only the relative size of the black population but its political organization and impact.

New England's substantial business displacement shows some unique characteristics. With a relatively low level of individual displacement and the smallest percentage of failures for displaced firms, New England clearly conforms to the model suggested earlier by Boston, with a heavy emphasis on downtown commercial renewal. The pattern of downtown renewal was by no means unique to Boston, or even Massachusetts. New Haven, Pawtucket, Stamford, Portland, Maine, and Burlington, Vermont, all mounted renewal efforts in their central business districts almost exclusively, and all show business discontinuance rates close to 20% or less.

The results in the Mountain and Pacific regions are strongly reminiscent of the character of renewal in Portland and San Francisco. For California, the individual displacement rate is 44% and the proportion of business discontinuances reaches 40%. The figures for Oregon are 60% and 42%. The pattern of the Far West is one of commercial renewal centered on the weakest and least politically powerful area—skid row and its surroundings.

Although the stage of development or the age of a locality's physical form may explain *part* of the variation in relocation and displacement, it is the local political environment that pushed one or another type of renewal to center stage, and provided local resources to a particular variant of urban revitalization. Thus, a policy of "Negro removal" in the South not only placed the burdens of displacement on a powerless group, but also promised the active support of the powerful. As Clarence Stone (1976) has ably documented for Atlanta, black neighborhoods could easily be replaced by large new civic centers, convention halls, and municipal stadiums. These developments won the favor of downtown business interests, for they not only replaced the unpleasantness of the slums but they also promised new traffic, shoppers, and workers for the downtown cores they surrounded. The approach and politics were much the same in the Far West, with the "bums and booze" that bordered the downtown fulfilling the role of target population. Only in New England is there a substantial difference in both politics and outcomes, with displacement falling largely on substantial business firms rather than on families and voters.

FEDERAL URBAN AID AND THE LESSONS OF RENEWAL

Contrary to the common view that the federal government has little capacity to shape local implementation of its programs, the impact of its redefinition of urban renewal goals in 1967 raises the possibility that when the federal government *can* clearly articulate its goals, it can elicit local response. Indeed, local agencies showed even higher levels of participation in the program (in terms of new projects initiated) after the imposition of federal constraints. This occurred despite the fact that the production of new housing for low- and moderate-income families and minority group members was hardly an objective shared by all community leaders or influential groups. The result was a national shift in the benefits of the renewal program, from an era before 1967 when the average project provided 47% of its new housing units to lower-income persons to an era in which low- and moderate-income families occupied over 65% of the new units.

The local response included a radical shift in the mode of operations of local renewal authorities. Bulldozer clearance gave way to housing rehabilitation and preservation. The exclusion of project residents from the planning process was replaced by representation on Project Area Committees and an increasing level of resident participation and influence.

The achievement of these purposes has been dependent on a number of factors, most of which were present in the last years of urban renewal. The existence of supportive local organizations, in the form of renewal agencies, provided a built-in political constituency in thousands of cities across the nation. These agencies, most with long histories of activity, not only had their own sources of local political support, but continuing needs for their own survival and growth. The need for a regular infusion of federal grant funds meant that local agencies would often adopt programs markedly at variance with their previous efforts when there was no practical, or more flexible, alternative.

The national political environment of the late 1960s also proved supportive of a new assertion of federal concerns and substantial local endorsement. This environment not only stiffened the backs of legislators and executives in asserting social goals, it also supported the poor and minority group members in their battles at the local level.

Finally, the renewal program could continue to capitalize on the local political need for "big projects" with their attendant symbolic and financial rewards. Renewal still provided the opportunity for the public display of models of the bright, future city, and the announcement of dramatic new local developments. It merely shifted their focus and reduced the degree of local discretion in choosing objectives and locations.

NOTES

1. Despite the short-term costs in lost tax revenues, government control of land is a useful means of guiding and shaping urban development. In many communities, the value of a renewal site became apparent only some years after its clearance.

2. The data on race and relocation are clouded by changes in categories and definitions. Until the mid-1960s information was only recorded for whites and nonwhites. For the balance of the decade, a tripartite division of whites, blacks, and other minorities was employed. Official figures for relocation activities during the 1970s include separate data for Spanish-surnamed persons.

3. The analysis is based on a population of 2,797 projects initiated during the twenty-five-year history of the renewal program. It includes traditional urban renewal projects as well as project areas under the Neighborhood Development Program. Each project's land use characteristics count equally in the average figures regardless of project size. This prevents a few very large projects from dominating the annual values. The residential rehabilitation measure is based on a slightly smaller group of projects (2,646) which excludes those areas with no residential land within their boundaries.

4. The analysis in Tables 4.3 and 4.4 is based on a total of 1,741 projects. This includes all projects initiated before January 1, 1968 in the continental United States.

5.

New England
Connecticut
Maine
Massachusetts
New Hampshire
Rhode Island
Vermont

Middle Atlantic
Delaware
New Jersey
New York
Pennsylvania

East North Central
Illinois
Indiana
Michigan
Ohio
Wisconsin

West North Central
Iowa
Kansas
Minnesota
Missouri
Nebraska
North Dakota
South Dakota

South
Alabama
Arkansas
Florida
Georgia
Louisana
Mississippi
North Carolina
South Carolina
Texas
Virginia

Border
Kentucky
Maryland
Oklahoma
Tennessee
Washington D.C.
West Virginia

Mountain
Arizona
Colorado
Idaho
Montana
Nevada
New Mexico
Utah
Wyoming

Pacific
California
Oregon
Washington

6. The relocation analysis is based on a total of 2,086 projects in execution as of June 30, 1971. The relocation and displacement figures reflect the *total* local or regional activity in all projects, and thus implicitly reflect the size of local renewal efforts. This provides the most accurate picture of the impact of urban renewal.

7. The relatively high business failure rate for the Yerba Buena project is largely a product of many firms (hotels, bars, restaurants) serving skid row which were not reestablished. For two views of that project, see Hartman (1974) and Wirt (1974: 295-305).

8. Any analysis of racial displacement by city and region is complicated by substantial differences in nonwhite population across the country. A more detailed examination of race, controlling for the proportion of nonwhites in a city's population as of 1960 and eliminating cities with large Hispanic populations, indicates that *relative* nonwhite displacement was lowest in the Border and New England regions, and highest in the South and East North Central areas. See Sanders (1977: 326-330).

REFERENCES

ANDERSON, M. (1964) The Federal Bulldozer. Cambridge, MA: MIT Press.

DIAMONSTEIN, B. (1978) Buildings Reborn. New York: Harper & Row.

FRIEDEN, B. and M. KAPLAN (1975) The Politics of Neglect. Cambridge, MA: MIT Press.

GANS, H. (1961) The Urban Villagers. New York: Free Press.

GREER, S. (1965) Urban Renewal and American Cities. Indianapolis: Bobbs Merrill.

HARTMAN, C. (1974) Yerba Buena. San Francisco: Glide Publications.

JOYNER, C. (1967) "Tucson: the eighth year of the seven-year itch," pp. 165-195 in L. Goodall (ed.) Urban Politics in the Southwest. Tempe: Institute of Public Administration, Arizona State University.

SANDERS, H. (1977) "The politics of city redevelopment: the federal Urban Renewal Program and American cities, 1949 to 1971." Unpublished Ph.D. dissertation, Harvard Univ.

STONE, C. (1976) Economic Growth and Neighborhood Discontent. Chapel Hill: Univ. of North Carolina Press.

U.S. Congressional Research Service (1973) The Central City Problem and Urban Renewal Policy. Washington, DC: U.S. Government Printing Office.

WIRT, F. (1974) Power in the City. Berkeley: Univ. of California Press.

5

The Impacts of Urban Aid

JOHN P. ROSS

□ OTHER CHAPTERS IN THIS VOLUME document an extensive history of direct federal efforts aimed at the redevelopment of our nation's largest cities. While the federal involvement itself is not new, the magnitude of that effort as measured by the size of direct federal grants-in-aid has substantially increased in recent years. The last decade has witnessed an enormous growth in the amount of this direct federal financial assistance. Federal grants-in-aid to cities increased by 585%—from $1.3 billion to almost $8.9 billion—between fiscal year (FY) 70 and FY 77. After discounting for inflation the growth has still been substantial. In real terms, federal aid to municipalities increased from $1.1 billion to $4.9 billion or by 345% over this period.

Individually, the programs responsible for this increase reflect a new, less restrictive attitude on the part of the federal government toward control over its aid to state and local governments. In the aggregate these grants may have had just the opposite effect. Increases of the magnitude suggested above in such a short period of time imply an increased centralization of our system of fiscal federalism and as a result a much tighter connection between the actions of the federal government and the municipal allocation of scarce resources. Municipal governments are now much more dependent on federal dollars to finance public services than they were in 1970. In FY 70 the average city received $0.07 from the federal government in direct aid for every $1.00 it raised in taxes. By FY 77 it

AUTHOR'S NOTE: *The author wishes to express his thanks to Phyllis Hedeman for providing valuable research assistance on this chapter. The opinions expressed here are those of the author and do not necessarily reflect the opinions or positions of the U.S. Department of Housing and Urban Development.*

received $0.24 for every $1.00 it raised in own-source general revenues. The largest 48 cities received an average of almost $0.48 for every $1.00 of own-source revenue in FY 77.

It is only fair to point out that the dependency relationship works in both directions. Not only are cities becoming increasingly dependent on federal aid to balance their budgets, but the federal government also is becoming increasingly dependent on the cities to carry out certain public services. Through the Comprehensive Employment and Training Act, for example, the federal government relies upon local governments to become employers of the low-skilled, hard-to-employ part of the labor force, a task which most local governments would not have imposed upon themselves without outside prompting.

The purpose of this chapter is to provide a discussion and analysis of the impacts of these grants-in-aid programs and a look at future changes which can be expected as a result of these substantial shifts in financial responsibility. The first section will examine the growth of grant-in-aid programs by concentrating on the forty-eight largest cities. The second section will look at the potential impacts of these programs and touch on a number of issues which result from the increasing reliance of cities on direct federal aid.

GROWTH IN DIRECT FEDERAL AID TO CITIES

The growth in direct federal aid to cities was most recently noted by Shannon and Ross (1976: 189-212) in an article documenting municipal dependency on both federal and state aid. The authors point out that from 1962 to 1975 all municipalities increased their dependency, measured as the ratio of aid to own-source revenues, at an average annual rate of growth of 10.4%. Large municipalities, those with over one million in population, increased their dependency even faster; their rate of increase amounted to over 13% per year.

There are at least two ways to measure the dependency of a place on direct federal aid: (1) aid per capita; or (2) aid per dollar of own-source revenues. The first measure shows the dependency of the people residing in a place on federal aid for the delivery of certain kinds of services. The second measure indicates the budgetary dependency of the local unit of government.

Table 5.1 presents the amount of aid per capita received by 48 of our largest cities for various years while Table 5.2 shows the rate of change in that aid since 1960. The range in per capita federal aid for 1978 is substantial, from $825.99 in Washington, DC and $254.57 per capita in

(text continued p. 131)

TABLE 5.1   Per Capita Federal Aid to the 48 Largest Cities,
1960, 1970, 1975, and 1978 (dollars)

	1960	1970	1975	1978[a]
Atlanta	0.67	8.15	88.40	189.29
Baltimore	2.58	33.92	127.65	212.55
Birmingham	1.69	3.82	64.64	122.92
Boston	0.01	26.09	104.85	196.85
Buffalo	12.90	8.27	78.21	211.58
Chicago	2.90	26.98	53.60	132.34
Cincinnati	4.39	50.06	148.31	201.44
Cleveland	3.61	4.46	74.72	98.85
Columbus	6.02	7.80	38.25	98.85
Dallas	b	3.54	30.66	51.69
Denver	0.99	21.87	102.20	164.13
Detroit	2.04	29.40	124.47	237.34
El Paso	9.79	4.01	39.04	116.80
Fort Worth	0.22	1.85	28.97	62.89
Honolulu	0.90	9.30	60.88	95.23
Houston	0.44	3.28	34.57	59.78
Indianapolis	0.06	1.57	50.71	106.06
Jacksonville	0.32	2.97	57.23	76.83
Kansas City	2.56	10.69	82.22	132.86
Long Beach	0.02	2.56	18.64	98.67
Los Angeles	0.66	2.02	42.18	114.93
Louisville	6.48	44.16	108.24	192.11
Memphis	2.07	2.76	37.65	71.21
Miami	2.43	0.01	22.71	146.18
Milwaukee	1.20	8.29	50.87	115.36
Minneapolis	0.05	8.32	39.38	153.71
Nashville-Davidson	32.16	18.80	60.80	85.37
Newark	0.05	6.40	46.01	252.01
New Orleans	3.39	4.83	81.59	232.35
New York	4.13	19.99	82.03	171.39
Norfolk	33.59	26.47	103.50	207.32
Oakland	2.34	25.46	141.34	193.19
Oklahoma City	6.37	12.30	51.08	101.80
Omaha	0.18	7.26	45.64	122.01
Philadelphia	5.29	15.77	72.05	185.85
Phoenix	1.35	11.65	54.99	123.56
Pittsburgh	2.06	20.10	55.24	206.23
Portland	b	16.89	76.90	171.27
St. Louis	0.86	8.08	59.97	208.92
St. Paul	0.04	3.58	33.86	140.14
San Antonio	0.50	2.63	43.83	89.71
San Diego	b	2.22	36.96	132.07
San Francisco	5.86	59.24	101.85	254.57
San Jose	7.79	4.81	22.41	96.59
Seattle	0.12	9.36	58.41	178.57
Toledo	4.76	8.27	52.53	136.16
Tulsa	0.09	13.82	78.66	72.18
Washington DC	77.59	326.39	167.88	825.99

a. 1976 population estimates.
b. Less than $0.01.
Source: Calculations based on Bureau of the Census, *City Finances*,
various years.

TABLE 5.2 Annual Average Rate of Change in Per Capita
Federal Aid for the 48 Largest Cities, 1969–1978
(percentages)

	1960–1970	1970–1975	1975–1978[a]
Atlanta	112	197	38
Baltimore	122	55	22
Birmingham	13	318	30
Boston	b	60	29
Buffalo	−4	169	57
Chicago	83	20	49
Cincinnati	104	39	12
Cleveland	2	315	44
Columbus	3	78	53
Dallas	b	153	23
Denver	211	73	20
Detroit	134	65	30
El Paso	−6	175	66
Fort Worth	74	293	39
Honolulu	93	111	19
Houston	65	191	24
Indianapolis	252	626	36
Jacksonville	83	365	11
Kansas City	32	134	21
Long Beach	b	126	143
Los Angeles	21	398	57
Louisville	58	29	26
Memphis	3	253	30
Miami	−10	b	181
Milwaukee	59	103	42
Minneapolis	b	75	97
Nashville-Davidson	−4	45	13
Newark	b	124	149
New Orleans	04	318	62
New York	38	62	36
Norfolk	−2	59	33
Oakland	99	91	12
Oklahoma City	9	63	33
Omaha	393	106	56
Philadelphia	20	71	53
Phoenix	76	74	42
Pittsburgh	88	35	91
Portland	b	71	41
St. Louis	84	128	83
St. Paul	885	169	105
San Antonio	43	313	35
San Diego	b	313	86
San Francisco	91	14	50
San Jose	−4	73	110

TABLE 5.2   Annual Average Rate of Change in Per Capita
Federal Aid for the 48 Largest Cities, 1969–1978
(percentages) (Cont)

	1960–1970	1970–1975	1975–1978[a]
Seattle	770	105	69
Toledo	7	107	53
Tulsa	b	94	–3
Washington DC	32	–10	130

a. 1976 population estimates.
b. Over 1.000% increase.
Source: Calculations based on Bureau of the Census, *City Finances*,
various years.

San Francisco to $51.69 per capita in Dallas. This compares to a range of
from less than a penny in Dallas, Portland, and San Diego to $77.59 in
Washington, DC and $33.59 per capita in Norfolk in 1960.

The average annual rate of change shown in Table 5.2 also has a
substantial range. From 1975 to 1978 per capita aid grew at between
181% per year in Miami and a minus 3% per year in Tulsa.

Federal aid per dollar of own-source revenues as shown in Table 5.3
indicates similar increases. In 1960, 22 of the 48 cities examined received a
penny or less in Federal aid for every dollar of own-source revenue. By
1978, 21 of these cities received $0.50 or more for every $1.00 they raised
themselves with Pittsburgh estimated to have received more than $1.00 for
every $1.00 of own-source revenue.

The average annual percentage growth in the ratio of direct federal aid
to own-source revenue varied from over 138% in Miami to a negative 9% in
Washington, DC, and a negative 5% in Tulsa from 1975 to 1978. Most of
the increase took place during the 1970 to 1975 period and was due at
least in part to the 1974-1975 recession which in some cases substantially
reduced cities' own-source revenue while at the same time federal aid was
being increased.

Such data document the importance of direct federal grants-in-aid to
these large cities. They also imply a substantial increase in the potential of
the federal government to directly influence the budgetary decisions of
these cities. The prospect of losing $0.50 for every $1.00 of own-source
revenue may be enough to induce most mayors to allocate their budgets in
ways which are acceptable to the donor government. It is this magnitude
of reliance on federal aid that inspires the use of the term "dependency."

Five programs are principally responsible for the dramatic increases in
aid. They are General Revenue Sharing (GRS); Community Development
Block Grants (CDBG); the Comprehensive Employment and Training Act

TABLE 5.3 Federal Aid Per Dollar of Own-Source General Revenues for the 48 Largest Cities 1960, 1970, 1975, and 1978 (dollars)

	1960	1970	1975	1978[a]
Atlanta	0.01	0.05	0.27	0.49
Baltimore	0.02	0.12	0.31	0.44
Birmingham	0.03	0.03	0.26	0.41
Boston	b	0.06	0.16	0.22
Buffalo	0.11	0.05	0.26	0.62
Chicago	0.04	0.19	0.22	0.48
Cincinnati	0.03	0.17	0.26	0.28
Cleveland	0.05	0.03	0.32	0.64
Columbus	0.10	0.07	0.20	0.45
Dallas	b	0.03	0.13	0.18
Denver	0.01	0.10	0.23	0.31
Detroit	0.02	0.15	0.42	0.66
El Paso	0.09	0.05	0.30	0.72
Forth Worth	b	0.02	0.16	0.28
Honolulu	0.01	0.06	0.27	0.36
Houston	0.01	0.03	0.17	0.23
Indianapolis	b	0.02	0.27	0.45
Jacksonville	b	0.03	0.27	0.28
Kansas City	0.03	0.06	0.25	0.23
Long Beach	b	0.01	0.06	0.26
Los Angeles	0.01	0.01	0.39	0.35
Louisville	0.06	0.22	0.20	0.80
Memphis	0.03	0.02	0.13	0.35
Miami	0.02	b	0.13	0.67
Milwaukee	0.01	0.07	0.27	0.56
Minneapolis	b	0.06	0.12	0.56
Nashville-Davidson	0.20	0.09	0.16	0.18
Newark	b	0.02	0.12	0.56
New Orleans	0.04	0.03	0.34	0.55
New York	0.02	0.04	0.10	0.16
Norfolk	0.32	0.13	0.30	0.58
Oakland	0.02	0.13	0.48	0.55
Oklahoma City	0.13	0.10	0.28	0.36
Omaha	b	0.08	0.28	0.64
Philadelphia	0.05	0.07	0.22	0.38
Phoenix	0.03	0.11	0.35	0.61
Pittsburgh	0.03	0.14	0.33	1.04
Portland	b	0.11	0.36	0.67
St. Louis	0.01	0.04	0.17	0.49
St. Paul	b	0.03	0.16	0.54
San Antonio	0.01	0.03	0.39	0.64
San Diego	b	0.02	0.24	0.67
San Francisco	0.04	0.14	0.16	0.33
San Jose	0.09	0.04	0.14	0.42
Seattle	b	0.07	0.21	0.49
Toledo	0.06	0.07	0.29	0.63
Tulsa	b	0.15	0.33	0.28
Washington DC	0.31	0.56	0.90	0.65

a. Estimated 1978 own-source general revenues.
b. Less than $0.01.
Source: Calculations based on Bureau of the Census, *City Finances*, various years.

(CETA), Titles II and VI; Anti-Recession Fiscal Assistance (AFRA); and the Local Public Works program (LPW).

Table 5.4 shows the per capita allocations of each of these programs for 1978. Again the range is quite substantial, particularly for individual programs. The total per capita amounts, however, in all cases are sizable. In FY 78 these five grants-in-aid programs provided more than 77% of direct federal aid to the forty-eight large cities.

These programs were all initiated before Carter came into office in 1976. Their operations reflect a combination of Nixon's "New Federalism," Congress's own "urban policy" and the Carter Administration's approach to urban aid which made only marginal changes in the programs.[1] In theory, they are a long way from the potentially highly targeted, tightly controlled categorical grants of the 1960s; however, in practice the size of these programs alone increases potential federal control over the allocation of resources by these cities.

GRS, the center-piece of Nixon's New Federalism, embodied a hands-off attitude on the part of the federal government toward the way state and local governments spend their new grants-in-aid resources. That new attitude was reflected in both the lack of restrictions placed on the use of program funds and the loosely defined program goals. One of the more important purposes was to redistribute the tax burden which must be borne for public services from the generally more regressive state and local government tax systems to the more progressive federal tax system. Residents of large cities who normally bear a heavier tax burden than their suburban counterparts were given some opportunity for tax relief. While the program was not directly aimed at large cities, its formula worked in such a way as to reward places with large population, poor residents, and high tax burdens. Older, declining urban areas most often have this set of characteristics and therefore tended to receive larger proportions of GRS funds.

GRS is scheduled for renewal in 1980. At present it appears that the renewal debate will concentrate on ways the program can be more highly targeted. There is much discussion about how this money can be restricted in a manner to force states to provide additional help to their urban areas. It is probable that neither the hands-off attitude nor the general types of goals so much a part of the original program will survive the battle over renewal.

The CDBG program and the CETA program are also creatures of Nixon's New Federalism. Both are block grant programs intended to incorporate sets of older categorical grants-in-aid into one system. In both cases, federal control over the program was reduced as a part of the consolidations. With each renewal, however, the formulas have been more

TABLE 5.4   Per Capita Federal Aid to Largest Cities by Program, 1978
(dollars)

	GRS	CDBG	LPW (Round II)	CETA (Titles II & VI)	ARFA	Total
Atlanta	17.60	32.69	16.91	66.36	4.90	133.56
Baltimore	33.21	34.15	23.67	36.76	11.50	127.79
Birmingham	29.51	38.61	11.08	32.74	8.95	120.89
Boston	37.87	39.84	25.83	52.26	8.83	174.82
Buffalo	18.59	52.23	34.18	66.94	9.78	181.73
Chicago	42.29	37.18	12.01	33.79	4.65	125.27
Cincinnati	24.95	41.45	25.67	46.33	7.83	145.79
Cleveland	25.11	55.18	15.26	38.82	6.29	140.66
Columbus	18.43	16.31	12.40	34.52	3.46	85.12
Dallas	17.84	17.56	–	14.44	–	49.84
Denver	26.92	24.54	17.76	29.38	5.22	103.82
Detroit	30.63	42.88	20.39	44.97	12.35	151.22
El Paso	19.02	21.93	18.05	39.39	9.71	108.10
Fort Worth	32.44	37.77	–	44.38	1.60	116.19
Honolulu	23.10	16.12	3.13	34.88	4.93	82.16
Houston	14.64	16.62	5.95	17.16	1.04	55.41
Indianapolis	20.24	15.03	14.92	36.30	2.07	88.56
Jacksonville	16.05	19.58	–	21.99	1.95	59.57
Kansas City	27.13	25.93	4.99	34.28	3.72	96.05
Long Beach	12.06	21.04	22.39	37.04	2.87	95.40
Los Angeles	17.00	18.18	16.69	43.21	4.93	100.01
Louisville	33.15	36.16	1.73	32.42	1.39	104.85
Memphis	17.48	22.93	–	16.47	1.10	57.98
Miami	23.63	28.59	21.49	58.24	8.41	140.36
Milwaukee	20.66	30.30	14.21	35.70	3.30	104.17
Minneapolis	21.36	46.68	19.88	43.69	1.40	133.01
Nashville-Davidson	21.34	19.77	–	16.28	–	53.39
Newark	38.98	51.15	42.92	82.10	22.52	227.67
New Orleans	36.23	34.45	16.36	34.38	11.17	132.59
New York	41.11	30.13	25.85	40.13	16.07	153.29
Norfolk	27.47	39.84	9.52	23.59	4.18	104.60
Oakland	17.53	34.22	33.64	65.31	8.59	159.29
Oklahoma City	19.51	20.10	17.78	24.66	.19	82.24
Omaha	14.91	14.09	48.15	21.02	.10	98.27
Philadelphia	28.90	35.03	30.46	37.95	11.84	144.18
Phoenix	13.81	14.60	22.58	49.81	1.48	102.28
Pittsburgh	27.44	51.20	36.17	44.02	8.47	167.30
Portland	27.15	27.62	32.06	46.83	6.17	139.83
St. Louis	26.82	61.35	29.05	46.45	10.40	174.07
St. Paul	22.75	54.65	18.15	39.42	1.08	136.05
San Antonio	13.84	23.25	10.63	27.42	3.15	78.29
San Diego	11.41	13.50	21.70	48.16	4.30	99.07
San Francisco	32.20	39.21	45.29	59.73	9.97	186.40
San Jose	10.55	11.05	17.08	31.71	3.01	73.40
Seattle	20.92	31.89	31.06	45.62	3.77	133.26
Toledo	15.30	21.79	22.76	38.96	2.98	101.79
Tulsa	20.16	18.98	–	15.27	.10	54.41
Washington DC	40.95	46.40	42.74	38.06	12.59	180.74

Source: Community Development Bock Grant Program Directory of Allocations for FY 78, September 1978.

finely tuned and for the CETA program new controls primarily aimed at preventing substitution have been imposed.[2]

Both ARFA and LPW were initiated by Congress and passed over President Ford's veto in response to the financial problems state and local governments were having because of the 1974-1975 recession (Advisory Commission on Intergovernmental Relations, 1978, 1979). The ARFA program is in effect a highly targeted general revenue sharing grant going primarily to places with high rates of unemployment. The LPW program was intended to provide additional funds to build new capital facilities. The funds were distributed by formula and again went primarily to places with high rates of unemployment.

As a part of his economic recovery program, Carter extended and expanded ARFA, LPW, and CETA. His administration attempted both to reduce in size and more precisely target all three of these programs as the economy recovered. With CETA Carter was successful; however, at the present time both ARFA and LPW appear to have failed in Congress.

In a general sense, these five grant-in-aid programs constitute a national urban policy supported by a substantial commitment of federal funds. The problem is that they were originally passed to solve individual crises rather than as an overall urban policy. As the crises subside, for example, and the economy recovers, the programs are allowed to terminate; whereas, if viewed as urban initiatives to encourage revitalization they might well be continued. For an urban policy to be successful, it might be argued that grant-in-aid programs which are passed must be justified on grounds of urban requirements rather than in relation to national economic needs or for other reasons.

THE IMPACTS OF AID

Given the kinds of increases documented above, one may well ask about the effect they are having on local areas. The Office of Management and Budget (OMB) asked that question in Circular A-116. It was signed by the President on August 16, 1978 and calls for an "Urban and Community Impact Analysis" to identify the likely effects of proposed major programs and policy initiatives on cities, counties, and other communities. A-116 requires that all *new* major federal budget initiatives be accompanied by an urban impact statement when the initiative is submitted to OMB.

The circular addresses all types of initiatives including new programs, changes in tax provisions, new regulations, and new regulatory authorities as well as federal grants-in-aid. It divides the types of impacts to be examined into five general categories: (1) employment, (2) population size; (3) income, (4) the fiscal condition of state and local governments, and (5) "other," including neighborhood stability, housing availability, quality of public services, etc. The first three of these potential effects may be classified as impacts on the socioeconomic conditions of the place—the conditions of the place as a population and business center. The fourth concerns the general financial condition of the state-local public sector of that place. The two issues are presumably fundamentally related since, to a large extent, the better a place is doing as a population and business center, the stronger is the fiscal condition of its government.

Circular A-116 requires an analysis of federal public policy choices but, as is often typical, neglects to explain how such an analysis is to be done. In a recent paper discussing Circular A-116, Norman Glickman (forthcoming) identifies some of the methodological issues which must be addressed. The more important of these issues include: *timing*—short-run vs. long-run, direct vs. indirect impacts, and the *type of spatial unit* to be analyzed—the region, the city, or the neighborhood.[3]

Traditionally, examinations of the impacts of grants-in-aid are concerned with three issues. The first concerns distributional impacts or targeting. Where should the money go and did it get there? The second deals with the fiscal impact of the money on the budgets of the recipient governments. How was the money used and how, if at all, did it reduce the fiscal stress on those governments? A third impact is economic. How did the money contribute to the redevelopment of the place in terms of such considerations as employment income, or population size? Unfortunately, present models of economic impacts are inadequate. Therefore, we will discuss only the first two impact issues and a third quite different one—the political impacts of federal grants.

DISTRIBUTIONAL IMPACTS

When the question of what types of places should get the most money—targeting—is posed, the most frequent response is that the money should go to places suffering from the greatest fiscal stress. The only real problem with this response is that we have no general agreement on a quantifiable definition of fiscal stress. In a recent publication Peggy Cuciti (1978) divides urban "need" or stress into three dimensions—social need, economic need, and fiscal need. The social need dimension relates to the characteristics of the residents, including, for example, their income levels, health problems and educational deficiencies. The economic needs of a place concern its growth and development. This dimension involves the movement of people, jobs, and businesses into and out of the city. The fiscal dimension views the tax, debt, and overall budget position of the local government. As Cuciti (1978: 6) points out, this need is often the result of a "mismatch between the need of the local population for public services and the resources available to the local government to pay for those services."

Even if there were general agreement that these were the basic dimensions of urban stress, three major issues directly related to targeting grant-in-aid programs would remain. The first and most straightforward of these issues concerns quantification of the three dimensions, i.e., how are these categories of urban health to be measured? At present all that we have are general proxies for each dimension and we do not know how to weigh the various proxies objectively so that they can be sensibly aggregated. For example, two generally used proxies of fiscal need are relative tax effort (relative to other surrounding or other like places) and relative debt burdens. The problem is how to combine these two measures into one component of an index of need.

The second problem concerns aggregation of the various dimensions of urban need. It is neither clear that such aggregation should be attempted nor, if it should be, how to do it. Some experts such as Cuciti argue that the three dimensions should be kept separate. Others such as Nathan and Adams (1976) have attempted to aggregate these dimensions into a single index of urban fiscal health; however, there is still much debate over the form such an index should take. For example, their index is multiplicative. The results of the aggregation might have been quite different had they used an additive form.[4]

Given the difficulties involved in answering the question of which places should receive federal grants-in-aid, what is usually done is to analyze the characteristics of the place that actually received the money and then attempt to match these findings with the original objectives of the program. In general, these kinds of studies examine one or more factors which appear in both the need index and the grant-in-aid formula for that particular program. Such a choice makes the entire analysis circular. For example, we know that GRS targets well (provides more money) to places with large populations (Ross, 1975). ARFA, LPW, and CETA all provide most of their money to places with the highest unemployment rates while CDBG provides most of its money to places with relatively high proportions of older housing. Thus, the Advisory Commission on Intergovernmental Relations (ACIR), found simple correlations of .84, .89, and .78 between the per capita allocations of ARFA, CETA, and LPW with their respective unemployment rates for the forty-eight largest cities (ACIR, 1978: 27).

On a more sophisticated level, Cuciti has attempted to rate each of these programs against all three of her definitions of urban need. Her results are shown in Table 5.5. The most interesting part of this table is the variation in responsiveness of the different programs to the three dimensions of need. According to these ratings, GRS has a low responsiveness to social need but is highly responsive to fiscal need. A part of this result comes from the fact that tax effort, a fiscal need proxy, is a factor in the GRS formula whereas none of the proxies which Cuciti used for social need is in this particular formula.[5]

What these kinds of results tend to indicate is that we can write formulas that will send money to places with whatever kinds of characteristics we want the places to have. Once the characteristics are decided upon, the political process inevitably spreads the money over larger areas than originally planned; nevertheless, once chosen, the characteristics are rewarded. The key problem is one of deciding what characteristics *should* be rewarded. The targeting issue is not so much a matter of getting the money to the right place as it is a problem of deciding what kinds of characteristics deserve federal aid—deciding which is the "right" place.

Each of the above considerations involve individual targeting issues for individual programs. In the aggregate these programs are relatively well

TABLE 5.5   Responsiveness of Grants Distribution to Differences in Social, Economic, and Fiscal Need, Based on a Sample of 45 Cities

Responsiveness	Social Need	Economic Need	Fiscal Need
High	ARFA CETA (Title I) CDBG (1980 projection) LPW CETA (Title II)	CDBG (1980 projection) ARFA LPW	GRS ARFA CDBG (1977 distribution) LPW
Moderate	CETA (Title VI)	CETA (Title I) GRS CETA (Title VI)	CETA (Title I)
Low	GRS CDBG (1977 distribution)	CETA (Title II) CDBG (1977 distribution)	CETA (Title VI) CETA (Title II)

Note: These rankings are based on Pearson correlation coefficients. All programs listed as being highly responsive had correlations of .5 or higher with measures of need. Moderate responsiveness means the correlation was between .4 and .5. Correlations in the low responsiveness group ranged from .028 to .399. The correlations between CETA (Titles II and VI) and fiscal need were statistically insignificant at the .05 level, suggesting that the grants distribution may be unrelated to differences in need.

Source: Cuciti (1978: xvi).

targeted to large cities. Large cities receive more direct federal aid per capita than do their surrounding suburbs or than do smaller cities. Looking at only one characteristic of need, ACIR (1978: 27) found a simple correlation of 0.6 between the total per capita aid provided by ARFA, CETA, and LPW and Nathan's urban conditions index for the forty-eight largest cities. This correlation coefficient indicates that the programs are targeted reasonably well among distressed and nondistressed large cities.

FISCAL IMPACT

Questions of fiscal impact concern the ways the grant-in-aid programs affect the financial health of the recipient government. The question itself is straightforward. The answers, however, appear to be quite complex and as yet we have not had much success in generating reliable estimates of fiscal impact.

Two problems pose a major share of the difficulty. The first concerns the question of substitution. When given additional grant-in-aid funds, a government may spend all of the new money plus what it had planned to spend of its own money plus in some cases even more of its own money. In this case, the grant stimulates own-source spending by the recipient government. On the other hand, the recipient government may spend the grant money and at the same time reduce its own-source expenditures by an equal or more than equal amount. In the latter case, grant monies are substituted for own-source expenditures. Total spending does not go up. The grant money is used for either tax reduction or increases in cash balances. Since dollars cannot be traced through the budget process, it is very hard to estimate what the government would have done had it not received the additional grants-in-aid.

Even if the question of substitution could be solved, we are still faced with a second question concerning how the budget decisions made by the government affect the fiscal health of the place. For example, suppose that we knew that the entire grant was substituted for local own-source spending—used for tax relief. What can we then say about the fiscal health of the place? The answer really is "very little," since we do not know much about the relationship between the way a jurisdiction allocates its resources and its overall fiscal position. Tax relief may actually encourage business and families to move back into the city—or at least stop moving out—and in turn increase the tax base of the jurisdiction. In the long run that decision may do more to improve the financial position of a locality than an increase in expenditures would.

We do have at least some theoretical notions on the substitution issue. In the case of a closed-end lump-sum transfer—the ARFA type of grant—

the impact on expenditure depends upon the local government's propensity to spend versus its propensity to reduce taxes or increase cash balances. Given the nature of the grant, the expenditure impacts should be less than that of either an open-end matching grant or a closed-end categorical grant. Using an econometric model, Gramlich and Galper estimated that after one quarter, a $1.00 lump-sum transfer grant should increase surpluses by $0.96 with only a $0.04 increase in expenditures or reduction in taxes. In the longer run that $1.00 increase in the lump-sum grant will increase expenditures by between $0.25 to $0.43 (Gramlich and Galper, 1973: 15-65).

While these kinds of estimates are available for single grant-in-aid programs, estimates on the potential effects of substitution for a group of programs are not yet available. It is not likely, however, that the aggregate substitution will be a simple summation of the substitution impacts of the individual programs. The grant-in-aid programs will interact with each other and change the expected responses of the local decision-makers.

AGGREGATE POLITICAL IMPACTS

The aggregate size of the increase in direct federal aid to cities raises a number of issues that are not directly program-related. These issues concern the behavioral responses of local jurisdictions to the total influx of federal dollars rather than to the individual program impacts such as those discussed above. There are at least four areas which need additional analysis. The first of these areas concerns the dependency relationship which such aid flows create. In addition, the higher level of government becomes more dependent on the local jurisdiction to deliver certain kinds of services. These public services vary in nature from those which are traditionally local in both financing and delivery to those which are generally considered to be national in scope.

For example, Titles II and VI of CETA use federal money to move local governments toward serving as employers of last resort. They also serve as employment agents of the federal government. In addition, these funds create a new level of government, that of the prime sponsor, which is most often composed of local governments and private, nonprofit agencies.

The rapid increase in CETA money to cities pointed to in much of the popular press has led to some unintended results. For example, it appears that at least at first, CETA money was often substituted for local funds to employ highly skilled workers. In other words, CETA was used as an additional source of general revenue sharing money (Palmer, 1977).

To the extent that such practices still occur, the federal intent of the program is subverted by local priorities. Local governments become active, but not necessarily accurate, agents of federal policy. A second example involves the ARFA program. One of the original justifications for this program was that the money should go to local governments so that they would spend more to speed economic recovery from a major recession. At the extreme, that kind of rationale moves local jurisdictions directly into the business of national economic stabilization—again, a business which a number of economists argue is not the place for local governments (ACIR, 1979).

A second concern involves the issue of control over resource allocations. At what point do local governments lose control over the allocation of their scarce resources? The standard assumption appears to be that there is a direct linear relationship between increased aid and loss of control over the aided function. However, given the creative ability of most local governments, it is unlikely that the relationship is that simple. It depends at least to some extent on the amount of local dollars originally allocated to that function. This problem is most severe when the function is a traditional local function such as education. When the function is a nontraditional one, local budget priorities are less likely to be severely affected.

The third area of concern involves uncertainty in the local budgetary process. Often, the greater the dependency on outside aid, the greater the uncertainty. That uncertainty in turn will affect the way local governments budget. Such a problem is particularly serious in those grant-in-aid programs with automatic termination dates. Those dates themselves may be sufficient to change the recipient government's behavior. The local allocations of GRS money provides an interesting example. The majority of the first-round allocations went to capital construction. As local governments accepted the fact that the money was going to continue, GRS funds were increasingly built into the operating budgets of local jurisdictions. Should GRS not be renewed in 1980, the budget disruptions could be quite substantial since most of the money now goes into operating budgets.

Finally there is the entire question of the states' role in the urban revitalization issue. Large increases in direct federal aid to cities have left the states almost entirely out of the revitalization process. The federal government, to some extent, has taken over the states' traditional role in helping its cities. Some recent evidence even indicates that as the federal role has increased, state aid to cities has declined (Olson and Stephens, 1979). One issue now is how to get the states back into the urban revitalization process.

FUTURE DEVELOPMENTS

Recent developments point toward a slowing if not to an absolute reversal in the rapid rate of growth of direct federal grants-in-aid. In addition, these trends point toward both increased targeting and more controls imposed by Washington on the way state and local governments can spend money. An increase in the aggregate amount of money flowing directly from Washington to cities gives the federal government much greater leveraging potential over the ways cities allocate their own funds. We predict that Washington will begin to use that greater leveraging ability. However, a sharp change in general economic conditions could reverse these trends and make the growth in aid even more rapid while loosening attempts at control. What we may actually be seeing is much less consistency in the growth of federal aid. It is possible that the federal response to the problems of the city will become much more dependent on cyclical changes in the national economy. Such a result would increase local governments' budgetary uncertainty, and perhaps force a much more active role on the states. Alternately, a demand for greater certainty on the part of local governments might force a major reexamination of federal-local relations.

NOTES

1. As the various contributions to this volume indicate, the Carter Administration has built upon the block grant approach while modifying it to target aid more directly to those cities and persons most in need. At the same time, its concern with stimulating and targeting economic reinvestment to central cities has resulted in the creation of such new programs as the Urban Development Action Grant and Neighborhood Strategy Area programs which resemble earlier project grant approaches. (On the former program, see the chapter by Gist.)

2. In most instances, these tighter controls are justified on the grounds that they are necessary to make the program more responsive to federal objectives. For an excellent case study of the CETA program which highlights this point, see Palmer (1977: 143-175).

3. A forthcoming volume edited by Norman Glickman contains a set of case studies which attempt to analyze the urban impacts of various major federal grant-in-aid programs. Because of the way the volume is organized, none of the papers attempts an analysis of the aggregate impacts of the programs.

4. The mathematical form of the index to some extent will determine the implied weights given to the different factors in the index.

5. It is interesting, although not surprising, to note that federal aid to cities and state aid to cities are often not highly correlated. One of the reasons is that the states and the federal government reward different types of place characteristics.

REFERENCES

Advisory Commission on Intergovernmental Relations (1978) Countercyclical Aid and Economic Stabilization. Washington, DC: U.S. Government Printing Office.
——— (1979) State-Local Finances in Recession and Inflation. Washington, DC: U.S. Government Printing Office.
CUCITI, P. (1978) "City need and the responsiveness of federal grants programs." Report for U.S. House of Representatives, Committee on Banking, Finance and Urban Affairs, Subcommittee on the City, 95th Cong., 2nd Sess. (August).
GLICKMAN, N. J. (forthcoming) "Methodological issues and prospects for urban impact analysis," in N. J. Glickman (ed.) The Urban Impacts of Federal Policies. Baltimore: Johns Hopkins Univ. Press.
GRAMLICH, E. M. and H. GALPER (1973) "State and local fiscal behavior and federal grant policy," pp. 15-65 in Brookings Papers on Economic Activity. Washington, DC: Brookings Institution.
NATHAN, R. P. and C. ADAMS (1976) "Understanding central city hardship." Political Science Quarterly 91 (Spring): 47-62.
OLSON, G. W. and G. R. STEPHENS (1979) 'Financing state and local services and the interlevel flow of funds in the United States, 1957 to 1977." Report to the National Science Foundation. Washington, DC.
PALMER, J. L. (1977) "Employment and training assistance," pp. 143-175 in J. A. Pechman (ed.) The 1978 Budget: Setting National Priorities. Washington, DC: Brookings Institution.
ROSS, J. P. (1975) "Alternative formulae for general revenue sharing: population based measures of need." Center for Urban and Regional Studies, Virginia Polytechnic Institute and State University, Blacksburg, VA. (June).
SHANNON, J. and J. ROSS (1976) "Cities: their increasing dependency on state and federal aid," pp. 189-212 in H. J. Bryce (ed.) Small Cities in Transition: The Dynamics of Growth and Decline. Cambridge, MA: Ballinger.

6

Defining a National Urban Policy: Bureaucratic Conflict and Shortfall

ERIC L. STOWE

☐ DURING HIS 1976 PRESIDENTIAL CAMPAIGN, Jimmy Carter called the adoption of a "comprehensive urban policy" one of his top domestic political priorities. However, by the time of his address to the National League of Cities in November, 1978, President Carter announced that the most urgent domestic problem—inflation—would make the 1980 budget "very, very tight," and that federal urban initiatives would have to be accordingly scaled down. On February 28, 1979, Carter withdrew two of his centerpieces for urban revitalization: the independent National Development Bank (originally Urban Development Bank or URBANK), and the Department of Development Assistance, which was to have been formed by consolidating the Department of Housing and Urban Development (HUD) with the Economic Development Administration (EDA) of the Department of Commerce (Washington Post, 1979a).[1]

Few would deny that the run-away inflation of the late 1970s was a significant factor in creating the gap between President Carter's early urban policy commitments and the withered urban package that remained in early 1979. However, it was not the major explanation for the urban

AUTHOR'S NOTE: *Many of the perspectives included in this chapter reflect my experiences as a NASPAA Faculty Fellow assigned to the U.S. Department of Housing and Urban Development and to the White House Conference on Balanced Growth and Economic Development, where I participated in several components of the urban policy process during 1977-1978. This chapter represents solely the views of the author and does not necessarily reflect the position of the University of North Carolina at Charlotte from which I was on academic leave as a NASPAA Fellow nor the position of my present employer, the American Chamber of Commerce Executives.*

policy "shortfall." The factors which led to the unfortunate demise of many of the urban legislative and administrative programs are as complex and varied as the causes of the condition of the South Bronx. However, one significant factor, and the one addressed in this analysis, was the urban policy *process* itself. This analysis will include three components: (1) a brief review of the historical policy context of the 1978 Urban Policy Report, (2) identification of the major organizational actors and the general sequence of events in the 1978 policy maze, and (3) an assessment of the most important organizational factors that accounted for the process outcome as of March, 1979.

THE HISTORICAL CONTEXT OF THE POLICY

The Urban Growth and New Community Development Act was passed by Congress in 1970. In large part, it was a response to two decades of criticism from academic and governmental circles that federal policies had a tremendous aggregate impact on metropolitan regions but that there was no systematic attempt to develop a comprehensive national urban policy. Hence the 1970 Act, generally known as Title VII, provided in Section 703 that the President should:

Transmit to Congress, during the month of February in every even numbered year beginning with 1972, a Report on Urban Growth for the preceding two calendar years which shall include
(1) information and statistics describing characteristics of urban growth and stabilization and identifying significant trends and developments;
(2) a summary of significant problems facing the United States as a result of urban growth trends and developments;
(3) an evaluation to meet such problems and to carry out the national urban growth policy;
(4) an assessment of the policies and structure of existing and proposed interstate planning and developments affecting such policy;
(5) a review of State, local and private policies, plans, and programs relevant to such policy;
(6) current and foreseeable needs in the areas served by policies, plans, and programs designed to carry out such policy, and the steps being taken to meet such needs; and
(7) recommendations for programs and policies for carrying out such policy, including such legislation and administrative actions as may be deemed necessary and desirable [Urban Growth and New Community Development Act, 1970].

While the intent of Congress seemed quite clear in identifying the requirements of Title VII, the response of the Nixon Administration was basically to downplay the issuance of a national growth report. The first report (White House, 1972), submitted reluctantly in 1972, devoted the majority of its text to asserting that there really was no urban crisis and that there should never be a national growth policy. Predictably, the 1972 report came under heavy criticism from Congress, from local and state governments, and from urban public interest groups. The objections emphasized three points: (1) the refusal to acknowledge the important impacts of the existing, uncoordinated complex of federal urban policies, (2) the almost total absence of analytic data and statistics in the report, and (3) the total avoidance of any effort to link the report with urban policy proposals being developed by the administration.

The 1974 report (White House, 1974), while slightly more positive in its orientation toward cities, was not a great leap forward. It was issued almost one year late, due to the predictable confusion surrounding the Watergate crisis and the presidential transition period. However, President Ford, again, did not concur with the notion that the report should be a central component of an integrated federal urban policy. Hence, the report, while marginally improved in terms of its analytic component, was still a pale image of a major policy document (Center for Responsive Technology, 1974).

Comparatively, the 1976 report was a further improvement (White House, 1976). President Ford ordered that the Domestic Council provide a review function, but that HUD share responsibility for drafting the basic document. The report was received as a significant "step in improving the process to deal with growth and development problems" (Franklin, 1976: 10). While many more urban data were submitted, three fundamental criticisms remained:

(1) The President's 1976 Report on National Growth and Development does not contain recommended national goals, policies or programs;
(2) the report lacks a theoretical framework needed to interpret the meaning of the analytic information; and
(3) the report fails to clearly articulate its assumptions and define its terms (Franklin, 1976: 11-13).

With the inauguration of the Carter Administration, the advocates of an aggressive and comprehensive urban policy had every reason to hope that the mission of the 1970 Act would finally be achieved. Indeed, Carter had proclaimed in a campaign address in New York City on September 5, 1976: "Our country has no urban policy or defined urban goals, and so we have floundered from one ineffective and uncoordinated program to another. . . . We need a coordinated urban policy from a federal government committed to develop a creative partnership with our cities" (Carter, 1976). Advocates viewed this presidential endorsement as the key missing link in the previous three Urban Growth Reports and, equally important, saw the opportunity for a massive new allocation of federal monies to assist the cities.

THE MAZE OF THE NATIONAL URBAN POLICY
PROCESS, 1977-1979

MAJOR ACTORS

The Carter urban policy process was initiated in a presidential memorandum to the cabinet on March 21, 1977, announcing that Patricia Roberts Harris, the newly appointed Secretary of HUD, would head an interdepartmental task force—the Urban and Regional Policy Group (URPG). The URPG was formed from six core cabinet level agencies: HUD, Commerce, Treasury, Health, Education and Welfare (HEW), Labor, and Transportation. Other important independent agencies, such as the Environmental Protection Agency, were assigned official liaison roles (Congressional Quarterly, 1978: 207).

Fundamentally, Carter's policy intent encompassed four broad dimensions: (1) the development of a comprehensive analysis of the causes of urban problems and how federal policy related to urban conditions and trends, (2) a total review of existing federal urban programs with the explicit intent of improving coordination and reducing duplication, (3) the submission of proposals for new federal action where needed, and (4) the formal reorganization of bureaucratic authority for urban policy and programs. The mandate for the first assignment was given exclusively to HUD and was expected to be released as the 1978 *Report on Urban Growth* and to provide the theoretical basis for the policy recommendations. The second and third objectives were assigned to the URPG under the leadership of HUD. The fourth mission involving the development of proposals for formal reorganization of the urban policy complex was given to the Office of Management and Budget (OMB).

Initially, Secretary Harris designated Assistant Secretary Donna Shalala, who headed the Office of Policy Development and Research (PD&R) within HUD, to serve as the acting chairperson of the URPG. Shalala also won the authority to oversee the writing of the 1978 *Report on Urban Growth*—a decision which reversed the 1976 assignment of that function to HUD's Office of Community Development and Planning (CPD). The designation of the same assistant secretary to both roles was apparently intended to assure continuity between the analytic element in the report and the policy review and development mission of the URPG.

Within the White House, presidential advisers Stuart Eizenstat and Jack Watson were assigned to oversee the entire urban policy process and to cultivate "input" from congressional committees and leadership, the governors and mayors, and important interest groups.

Yet another organizational actor was the White House Conference on Balanced Growth and Economic Development, which had been mandated by the Public Works and Economic Development Act amendments of 1976. The White House Conference was chaired by West Virginia Governor John D. Rockefeller, IV, and staffed through the Department of Commerce under Secretary Juanita Kreps.

All of these efforts were consistent with President Carter's public commitments to an open administration, governmental reorganization and a comprehensive review of federal urban policy.

Public administrators and political scientists will long debate the nature of Carter's presidential leadership. However, as early as the summer of 1977, there were warning signs that the administration was unable to deal with the complexity of its own urban policy process, not to mention the range and depth of policy disagreements, the intensity of bureaucratic rivalries, and the built-in tensions generated by the expectations for simultaneous departmental reorganization, public participation as well as comprehensive policy review.

SEQUENCE OF MAJOR EVENTS

Any attempt to track all of the public statements, the private debates and the bureaucratic struggles is, of course, futile. The rivalries within HUD were hydra-headed and the process diffuse. Interagency conflicts were Byzantine. Nevertheless, the basic sequence of events can be summarized, and examples of organizational behavior can be examined as a way of identifying some of the critical dimensions of the policy shortfall.

March, 1977–Issuance of Executive Order #11297 Creating the Urban and Regional Policy Group. As with most new administrations, February and March are months in which a flurry of executive orders are issued to delineate "new directions." Unfortunately, new administrations typically underestimate the start-up time and the coercive power necessary for actual implementation. The URPG mission, moreover, went several steps beyond the scope of the normal executive order. It required coordinating a group of mutually jealous and competitive federal bureaucracies headed by political appointees who, for the most part, had never managed large organizations. It was mandated to conduct a comprehensive review of all federal programs which had major impacts on urban America, and to produce a coordinated policy package with attached cost analyses by mid-fall, 1977—in time for the preparation of the next budget—a Herculean charge by any standard. The probability of interagency cooperation was reduced still further by the parallel assignment given to OMB to examine the option of merging HUD and EDA (from the Department of Commerce) in order to assemble urban programs under a single umbrella. The possibility of such a major reorganization meant an extremely high level of uncertainty and placed the two most influential urban departments and their attached congressional and interest group supporters in defensive postures.

May 23-June 24, 1977—HUD Citizen Forums. In order to fulfill the presidential commitment to an "open policy process," Assistant Secretary Robert Embry, who headed CPD, was directed to sponsor a series of Citizen Forums in the ten HUD regional office cities. However, observers noted that these "town meetings" turned out to be more of a "government road show." They traveled with the predictable cast of federal staff and attracted local politicians, and a camp following of powerful interest group representatives who tracked the sessions from city to city. Few ordinary citizens even knew about the forums, much less were able to attend (Wall Street Journal, 1977).

Mid-Summer, 1977—The URPG Sessions and the Urban Policy Report. By mid-summer, 1977, the URPG had held only a few preliminary meetings which concentrated on organizational arrangements. Also, by July, the team which was to draft the 1978 *Urban Growth Report,* now retitled the *Urban Policy Report,* was just being assembled under the direction of a special assistant to the Assistant Secretary of PD&R. That team was drawn almost exclusively from recently employed academics on one-year assignments under the Intergovernmental Personnel Act.

September-October, 1977—The URPG and Urban Policy Report. In August, 1977, Vernon Jordan of the Urban League issued a blistering attack on URPG "inaction." In response, Secretary Harris felt the need to demonstrate movement within the URPG, and apparently acting on the assumption that leadership was the problem, appointed Assistant Secretary Embry as the new chair *pro tempore.* By October, the URPG was circulating draft proposals for revised and new urban legislative initiatives. However, to the team writing the *Urban Policy Report,* which was still assigned to PD&R, October brought the realization that, at best, their draft would be issued only simultaneously with the policy recommendations of the URPG, and, at worst, their analytic document would bear little resemblance to the URPG policy emphases. In short, the two documents were slipping out of synchronization (New York Times, 1977a).

November, 1977–"Cities and People in Distress." On November 6, 1977, a hastily assembled list of policy proposals was issued by the URPG under the title, *Cities and People in Distress.* The goal of achieving "something by November" was realized only by drawing upon a massive commitment of time and energy by a HUD consultant who consolidated the various "wish lists" of participating agencies into a single document (New York Times, 1977b). The keystones to the preliminary legislative/administrative set of recommendations were stitched into a "seamless web" of five policy clusters summarized in the following very general goals:

(1) create ample jobs for the urban poor, blacks, and other minorities and strengthen local economies;
(2) reduce urban fiscal and social service disparities;
(3) commit federal action to revitalize the physical environment of urban areas and their neighborhoods;
(4) strengthen the capacity of all levels of government and neighborhood groups to meet the needs of people and cities in distress; and
(5) expand freedom of choice and equity in urban areas (President's Urban and Regional Policy Group, 1977).

The final result of a competitive/collegial agenda-building process under extreme time pressure was basically a smorgasbord of proposals for the extension of existing programs. There was little effort to rank the projects among the various contributing agencies and almost no attention paid to combining or eliminating programs. The process, in fact, was remarkably similar to the agenda-building process of the Model Cities era, and, moreover, was staffed by some of the key participants of the prior decade's efforts.

The intensity of interagency competition was revealed in a detailed memorandum which was highly critical of *Cities and People in Distress* and which was sent to the White House by a senior appointee in the Department of Commerce. The Commerce draft specifically charged that the URPG Report provided "no basis for setting priorities or making choices among proposals" and that it "lacks a rationale for providing special aid for distressed cities and their residents" (Washington Post, 1977).

December, 1977–Carter Review of the URPG Package. A much reported and many-versioned meeting between the President and his chief urban policy advisers on December 14, 1977, resulted in what the *Washington Post* called a desperate attempt to "untangle the urban policy."

Carter, true to his administrative style, had expected an overall policy framework within which he could check off acceptable programs, strike unacceptable ones, and mark up the attached budgets. When he was presented with only a program list containing the most elementary policy umbrella and few cost figures, the President "requested" a detailed staff memorandum outlining the basic urban policy objectives (New York Times, 1977c; Congressional Quarterly, 1978: 21).

December 23, 1977-January 25, 1978-Joint Harris-Carter Memorandum. By December 23, 1977, Secretary Harris and presidential assistant Eizenstat, with around-the-clock efforts from their staffs, had drafted a "decision memorandum" for the President. Under heavy public pressure from urban and black groups since August, and anxious to show movement prior to the upcoming White House Conference on Balanced Growth, Carter initiated the memorandum with the admonition that he expected a complete policy-by-policy review to be conducted prior to the scheduled date of his urban policy announcement in March 1978.

January 29-February 2, 1978-The White House Conference on Balanced Growth and Economic Development. The White House Conference was particularly important because it offered a forum which could have been used to attack the meandering urban policy in the same tone as the August 1977 barrage by Vernon Jordan. While the conference was not formally linked to either the URPG activity or to the National Urban Policy Report, its staff was supported by the Department of Commerce. The Conference was composed of 500 citizens who had been selected by their state governors and by the White House to reflect a cross-section of interested local elites. Despite efforts to ward off divisive issues, the conference did raise a number of criticisms and concerns particularly relating to the Sunbelt-Snowbelt controversy and the role of citizen involvement in the policy process (White House Conference on Balanced Growth and Economic Development, 1978).

February-March, 1978-Base Program Review. Following the direction of the January presidential sign-off, the URPG agencies launched a base program review, most of which was assembled in a final frantic week of activity within HUD's Office of CPD in late March. Performance of this assignment demonstrated the difficulty that agencies face in objectively evaluating their major programs and their extreme reluctance to delegate that authority to a sister agency. The result was an "acceptable" program review with a policy tilt toward "distressed" cities (Congressional Quarterly, 1978: 21).

March 15-28, 1978-Urban Policy Release. Carter received his first URPG briefing on the base program review and the final policy proposals on March 15. Less than one week later, the voluminous written analyses were forwarded to the White House with a list of seventy policy options for presidential action. Under pressure to get an urban policy "on the street" before an extended overseas trip, Carter met with a few key advisers and decided which programs in the multibillion dollar complex would become elements of his urban package and which would not, all within a single weekend. In a hastily called congressional/press briefing session on Easter Monday, 1978, the President issued the final draft of the URPG report—*A New Partnership to Conserve America's Communities: A National Urban Policy.* [2] This title immediately induced confusion, for it is important to stress that this document was *not* the National Urban Policy Report—the analytic report to Congress—which was not issued until August 15, 1978 (*Congressional Quarterly,* 1978: 21-22).

June 16, 1978—Status Report on the Urban Policy. As an interim action, the White House released *A New Partnership To Conserve America's Communities: A Status Report on the President's Urban Policy* on June 16, 1978. This document was intended to fend off criticism about the slow progress toward issuance of the analytic report. The initial draft of the analytic report had been delivered to Assistant Secretary Shalala in early October 1977 and was then circulated extensively through PD&R for comments and revisions. By December, the anxiety concerning its compatibility with the URPG Report motivated CPD Assistant Secretary Embry to commission his own version of the Urban Policy Report. The Academy for Contemporary Problems was asked to take the PD&R draft and meld it with the URPG text. The hybrid document was delivered to HUD in January, 1978 but was still being rewritten in June.

August 16, 1978—Issuance of the National Urban Policy Report. The President's National Urban Policy Report—1978 was released at a press conference on August 16, 1978 (U.S. Department of Housing and Urban Development, 1978b). However, most of the President's remarks were devoted to four executive orders. [3] The announcement of the report was so understated that the coverage in the next morning's *Washington Post* made absolutely no mention of it. Compared to the 1972, 1974, and 1976 reports to Congress, it did meet the criticism of linking programs to policy, if only in a post hoc fashion. However, it was not a significant improvement in the breadth and power of previous analyses. Moreover, it suffered the ignominious fate of being issued long after the critical policy decisions had been made. The rational planning model, which Carter had initially envisioned, had been violated many times, but in no other instance was the disjunction quite so stark (Washington Post, 1978).

February 28, 1979—Withdrawal of the Keystone Proposals for Urban Agency Reorganization and URBANK. Two of the few major policy initiatives in the urban policy package were: (1) a proposal to merge EDA under HUD to create a new Department of Development Assistance, and (2) a proposal for URBANK. Both items encountered a great deal of congressional skepticism and were formally withdrawn by the Administration.

SUMMARY

In retrospect, the urban policy had never achieved the status of a coherent package. Those pieces which involved minor extensions of existing programs or only administrative adjustments fared well, for example the Comprehensive Employment and Training Act (CETA) program and the reallocation of EPA funds from peripheral urban to center city priority. However, actions which required significant changes in agency behavior, for example the commitment to spend federal purchasing dollars in depressed cities, have met with only marginal success. Initiatives which required close interagency cooperation, such as URBANK that was to have been cochaired by a triumvirate of secretaries from Treasury, Commerce, and HUD, tended to generate internecine conflict. Proposals which necessitated wholesale reorganization fared particularly poorly. Part of the demise of these proposals reflected a long-standing distrust of the competence of HUD, but the withdrawal also signaled a more basic concern that resulted from the fact that the programs had simply been developed too hastily in the press of trying to fulfill the administration's promise of reorganization (Washington Post, 1979b). (See the Appendix for details on the disposition of the major proposals.)

POLICY CONTEXT: RATIONAL MODEL VS. ORGANIZATIONAL REALITIES

In order to assess the causes of the urban policy shortfall, one must examine the realities of organizational management within which the process occurred. This analysis is limited to the administrative aspects only and does not address the legislative review.

Most planning theory is premised on a rational model of decision-making. As the congressional hearings on the 1970 Housing Act show, these rational assumptions permeated the mandate for the biennial Urban Growth Report. President Carter's technical background and leadership style also lean distinctly toward rationalistic assumptions. However, many of the key elements of the rational model did not operate in the context of the 1978 urban policy process, particularly in relation to: (1) the nature of the problem, (2) unitary control, (3) management skill, and (4) time.

PROBLEM ASSUMPTION

The first missing condition is the expectation that the "problem" is known and is definable. In the case of urban policy, however, once the discussion goes much beyond the level of "cities in crisis," there is great debate about the real condition of urban America and the measureable influence of federal policies. A related dimension of particular salience is the expectation that problem definition and solution selection are relatively value-free processes which are structured only by considerations of effectiveness, efficiency and the "best interests" of the nation. In the real world of organizational behavior, however, problem definition and solution preference are heavily influenced by personal experience, political loyalties, and bureaucratic predisposition. To use Allison's phrase, "where one stands depends upon where one sits."

And, as Wilensky (1967) notes, organizations typically use information and analysis as weapons. Of special importance are those analytic dimensions which are viewed as politically volatile or contrary to elite preference. One example in the 1978 policy process was the assumption made in the report that dense core cities are inherently energy efficient. Despite the glib media portrayals of the city as an energy haven, what little quantitative research has been completed paints a less than sanguine image, almost unanimously predicts a series of hard societal choices, and imposes a number of severe caveats on prognoses of energy efficiency in urban areas (Skidmore et al., 1977).

A related example, largely ignored because of the perception that it was incompatible with the goal of targeting, is scale and diversity of economic decentralization to "fringe" and "hinterland" areas. Muller (1976) and other economists and geographers argue that the entire concept of "city" must be reexamined. If the definition of "urban" is the intensity of economic activity with a second level including the frequency of social interchange, then graphing the present situation increasingly reveals a multinodal city. Such a redefinition could raise important questions concerning the probable effectiveness of policies targeted to central city revitalization without concomitant attention to the "outer urbanization." Critics point out that such a perspective could lead to a much different analytic outcome, and therefore, to different policy recommendations (Muller, 1976; Black, 1978).

Clearly, the urban policy "game" began with many of the preferred outcomes preselected. For example, the formal exclusion of the Department of Agriculture from the URPG, despite its huge Farmers Home Administration housing program and its influence on the peripheral growth of metropolitan regions, revealed that certain positions were selected out a priori, HUD came into the process dominating the action channel. The URPG chairperson was a HUD appointee, the support staff was primarily from HUD, and the analytic biennial report was being written within HUD. The agency had certain preferred policy outcomes due not only to its bureaucratic program inertia toward certain categories of solutions, e.g., public housing and central city renewal, but also a strong political tilt toward the problems of the older core cities, particularly those in the eastern megalopolis. Hence, the first barrage of slogans, most notably "distressed cities," which HUD had previously defined by carefully weighting its Community Development Block Grant formula to reward cities with large stocks of pre-1940 housing, set the tone. Other agencies quickly replied with their own slogans, EDA with "rapid growth stress" reflecting its concerns for the smaller and growing metropolitan areas and HEW/Labor's with "people in distress," as opposed to HUD's "places in distress."

It is not the intent of this analysis to evaluate the "rightness" of the various competing policy banners. However, they illustrate that the congressional/presidential expectations for the best possible analysis of the urban system, which could then be used to structure the necessary administrative and legislative debate, *did not* and *could not* result from a process which degenerated into a competition of policy slogans and positions.[4]

UNITARY ASSUMPTION

The second assumption of rational decision-making that did not operate is that of the problem-solver as unitary actor—that is to say, the organization responds to the problem in its environment with a single, constant locus of decision-making authority and is internally cohesive in its search for solutions. Even to approach the primary goals of the URPG charge— identifying the critical dimensions of the urban condition, specifying those federal activities which impinge on that condition, synchronizing and consolidating existing programs, and proposing an integrated set of policy initiatives—would have required a powerful body beyond the direct control of any of the participating agencies. This problem was partially anticipated in the 1970 Act when Congress suggested that the responsibility be assigned to the President's Domestic Council. Needless to say, the unitary rational ideal is operationally impossible. However, the nature of the procedures adopted by the URPG—a factionated unit with fractionalized power—was almost the polar opposite of the assumptions of control spelled out in the presidential charge.

In the turf-conscious world of the federal bureaucracy, placing any cabinet secretary in a position to review other agency programs, much less to evaluate their effectiveness and potentially consolidate, transfer or eliminate them, raises immediate storm warnings. Accordingly, the wary environment of the early URPG meetings produced only superficial commitments to unity. Moreover, the perceived threat of HUD domination was muddled when Secretary Harris designated an assistant secretary as her surrogate. Understandably, Harris, facing the numerous substantive and less substantive demands on a new cabinet officer, felt the need to delegate the responsibility. However, this action was also a subtle signal that the URPG was not really going to require cabinet-level participation, and, over the first several months, the cabinet representatives cautiously substituted a succession of deputies and special assistants to "represent the agency" at the meetings. Organizationally, the designation of subordinates to handle the actual workings of such massive assignments is quite predictable. The time of cabinet officers is much too limited to be absorbed in this type of process. It is even logical to assign persons with the day-to-day operational program knowledge to carry on such discussions. However, an unintended consequence is often a successive diminution of the visibility and the commitment of the larger agency to the mission and a parallel increase in overcautious behavior. Resultant undermining of any type of comprehensive review can also reflect the tandem fears that (a) "This is only a symbolic effort, and if I devote too much of my time here, I'll slip out of the mainstream"; or (b) "This is a serious effort in which we could lose much of our agency domain, so let's downplay it and see what happens." The end result is the same and it afflicted the early URPG which was characterized by symbolic activity with periodic intervals of turf-defensive fratricide.

MANAGEMENT SKILL

A third deviation from rationalistic assumptions involves the skill of decision-makers in directing their own complex organizations. As Hugh Heclo has eloquently suggested, few newly appointed cabinet and subcabinet officials come to Washington with much experience in managing sprawling political bureaucracies; few, in fact, have had responsibility for managing any large organization.

> The overall picture is one with a large number of supposed political controllers over the bureaucracy—transient, structurally divided, largely unknown to each other and backed by a welter of individual patrons and supporters. Held vaguely responsible for the actions of the government's huge organization, these national political executives are too plentiful and have too many diverse interests to coordinate themselves [Heclo, 1977: 242].

Unfortunately, the learning process for political appointees necessarily absorbs at least their first year in office. In part, the explanation of the 1978 urban policy process shortfall relates to the "decision haze" that inevitably envelops any new appointee. The early Carter Administration was particularly vulnerable because of the prevailing doctrine that the source of governmental inefficiency and ineffectiveness was located in the vast middle levels of the bureaucracy. The level of estrangement is even higher when the change of administration also involves a change in presidential party. The political appointees are insecure around top-level bureaucrats who survived or thrived under the "rascals" just thrown out. The mutual wariness at the beginning of the Carter Administration was at an even greater level, in large part due to the rhetoric used to anticipate promotion of major civil service reforms. Suddenly, the villains of inadequate federal policy were not the Republicans, but the ensconced civil servants. Consequently, large numbers of extremely knowledgable urban experts, who constitute much of the organizational memory, were functionally excluded or deliberately withdrew during the critical early stages of the URPG process.

TIME

Yet another rational model assumption which did not operate was sufficient time. As numerous decision theorists have illustrated, complex policy decisions which occur under extreme time pressures are likely to induce simplistic views of the problem and an overreliance on standardized solutions (Allison, 1971: 168; Cohen et al., 1972: 9-11). There is probably no more pertinent example of the negative results of a desperate push to deliver policy before the close of the "decision window" than the submission of the initial URPG list to the President in November 1977. Even with a multitude of analysts, no one could have realistically expected a coherent, integrated program review between the real start-up time of the URPG in September, and the initial release of *Cities and People in Distress* in November 1977. The predictable litany of "lets do more, only better" agency lists were the best product that any loose collectivity of large organizations could issue under extreme time and uncertainty pressures.

The conceptual haze, the competing slogans, the fractionalized power, the factionated rivalries, the managerial inexperience, and the extreme time pressures are all interactive factors which inexorably forced a superficial problem analysis, an artificial reduction of alternative solutions, and a hurried rearrangement of old policy repertoires under new labels; in short, they induced a phenomenon which frequently occurs in new administrations and which Graham Allison terms transitional error (Allison, 1971: 272).[5]

LESSONS FROM THE 1978 URBAN POLICY PROCESS

There are at least six salient points to be considered from the 1978 urban policy process. First, the initial responsibility for overseeing both the analytic and the policy formulation processes in such a process should be focused outside the bureaucracies themselves. Endemic rivalries simply overwhelm such processes without firm control from the White House. Interagency coordinating groups rarely escape the policy commitments of their agencies, and the political loyalties of their participants. The poor record of presidential commissions makes this option equally undesirable. Probably the only workable alternative is to assign the major responsibility to the Executive Office of the President. This is especially important since a truly encompassing urban policy process necessarily enters the sensitive zone of a national development or national growth policy. It is doubtful that establishing another, larger, body such as the Interagency Coordinating Council that replaced the URPG under Jack Watson's control, can perform much differently from its predecessor.

Second, even informal interagency cooperation will be severely inhibited by a simultaneous move to consolidate bureaucracies. Organization theory has long identified the uncertainty conditions which lead to defensive posturing. No higher level of uncertainty exists than threats of reorganization. Even if pursued separately, the goals of a collegial policy review process and governmental restructuring are difficult; pursued in tandem, they can engender paralysis.

Third, the time expectations for the long-overdue reanalysis of the urban condition and the necessary program consolidation need to be greatly extended. The time pressures perceived by a new administration are totally comprehensible, but they can have a detrimental impact on comprehensive analysis. Time is another reason to locate the coordinating responsibility in an independent unit because cabinet secretaries simply do not have sufficient time. Most observers of the federal decision-making process are painfully aware of how many critical choices are made by harried political appointees in thirty-second "briefings." The time requirements built into the 1970 Act for the submission of the report to Congress in even numbered years also insures havoc during transitions between administrations.

Fourth, despite the appealing logic of linking the analytic with the formulation process, the analysts need protection, however imperfect, from the powerful policy preferences of the bureaucracies. A genuine bedrock analysis simply cannot emerge if the report has to conform to the current buzzwords or if it is prevented from examining heretical policy options. Few observers would be so naive as to believe that all policy/political pressure could be removed, nor would they deem it desirable. But, even a modicum of insulation could enhance the quality of evaluation. Somewhere between "urban revitalization," the "new urban system," and "rural renaissance," for example, may lie the seeds of a comprehensive development policy.

Fifth, as several scholars have recently commented, considerably more attention has to be devoted to the incredible challenge of managing the managers. The stress of the current civil service reforms on the "flabby middle" has yet to be tied to a reexamination of the processes for selecting and overseeing the multiplicity of political appointees who, in fact, already exercise tremendous influence over which policy options emerge or are buried (Sundquist, 1979; Beam, 1978; Thayer, 1978).

Sixth, the selective use of information and the intense debate over policy priorities did not, of course, simply reflect mindless bureaucratic rivalries; they were also symptoms of the very real politically redistributive nature of federal urban programs. The definition of "distress," for example, has essentially been a dispute over political choices which has produced both winners and losers. Assuming the further reduction of federal allocations to urban areas, the intensity of this "data war," and the jockeying for political benefits associated with it will undoubtedly increase, making it imperative that the President be able to obtain more accurate and unbiased evaluations of urban conditions.

In sum, the overreliance of the urban policy process on a rational policy model, its failure to compute the intensity of bureaucratic rivalries and the importance of hidden personal agendas, and its error in endorsing regionally and intergovernmentally biased definitions of "urban distress" all combined to produce a significant policy shortfall in 1978.

APPENDIX: 1978 URBAN POLICY PROPOSALS: RECORD OF CONGRESSIONAL ACTION

Proposals Endorsed

Neighborhood Self Help Fund
$15 million HUD project to support neighborhood and voluntary organizations

Air Quality Planning
$25 million EPA program for localities to coordinate environmental and transportation planning

Solid Waste Recovery Planning
$15 million EPA project to localities for feasibility studies

Proposals Rejected or Deferred

National Development Bank
$2.2 billion in loan guarantees for companies locating in distressed areas; $550 million in Title IX and urban development action grants; $1 billion to set up a secondary mortgage market for commercial loans

Labor Intensive Public Works
$1 billion to rehabilitate and renovate public facilities, with 50% employment from the hard-core unemployed

APPENDIX (Cont)

Livable Cities
$20 million for a community arts program to be administered by HUD and the National Endowment for the Arts

Targeted Employment Tax Credits
$1.5 billion per year revenue loss for tax credits to employers who hire young CETA trained workers

Urban Parks
$150 million to local governments

Section 312 Rehabilitation
$150 million supplemental to HUD beyond $95 million initially proposed

Transportation Connected Development
$200 million for an Urban Mass Transit program for transit facilities (adopted with major congressional revisions)

Supplementary Fiscal Assistance
$1 billion to replace existing countercyclical program

State Incentive Grants
$200 million to redirect state resources toward distressed areas

Urban Volunteer Corps
$40 million technical assistance program administered by ACTION

Neighborhood Crime Prevention
$10 million to be administered by LEAA and ACTION

Investment Tax Credit
A 15% differential would be allowed for firms locating in distressed areas

Source: *National Voter,* 28 (Winter, 1979).

NOTES

1. The decision not to submit the proposal for reorganizing HUD and EDA into a Department of Development Assistance offered the opportunity to assign most of the proposed National Development Bank authority and monies to EDA. This option appeared likely to secure congressional approval, although it was regarded as a serious blow to HUD and a reaffirmation of EDA's role within Commerce.

2. The *New Partnership* report (President's Urban and Regional Policy Group, 1978) did not include a list of specific programs but rather only a series of broad principles. The action list remained quite similar to the one proposed in the November 1977 *Cities and People in Distress.* As Donovan illustrates in his study of the proposal generation process employed in the 1964 War on Poverty, agency-by-agency solicitation inevitably produces a laundry list with no coherent policy framework (Donovan, 1967).

3. The four executive orders included:

(a) creation of yet another Interagency Coordinating Council "for coordination and implementation of Federal urban and regional policy" to be composed of fifteen agencies. Essentially, the Council, headed by presidential advisor Jack Watson, would serve as an enlarged URPG;

(b)   assigning responsibility for federal space management to the Administrator of the General Services Administration (GSA). This order specifically directs the GSA to tilt toward core locations for federal facilities but allows considerable agency-by-agency discretion;

(c)   assigning to the GSA a responsibility to encourage federal procurement in labor surplus areas. In March 1979, however, the Northeast-Midwest Institute, the research unit of the coalition of 213 Northeastern and Midwestern Congress members, asserted that few agencies were complying; and

(d)   an order to the Director of OMB to institute an "urban and community impact analysis" system through which every major federal policy would have to be screened to insure that the urban impacts were identified and evaluated. This approach is quite similar to the ill-fated A-95 review process, but on a much larger scale, and with even less certainty of what such a review process was intended to examine.

4. Wilensky (1967) makes the interesting point that organizational propaganda, which frequently takes the form of slogans, while directed externally, has a pathological tendency to be most effective on the source itself. Slogans become so integral to the decision-making process that they subtly act as constraints on organizational behavior.

5. In addition to Allison's and Heclo's observations, one additional theorist has shed important light on an intense policy process. Irving Janis's concept of "groupthink" (1972) is important in understanding how a working group artificially reduces options to a few "tried and true" program approaches. This behavioral condition is a plausible explanation of the inertia of old ideas. It led to the frustrated reaction of a White House adviser, who commented that the urban policy proposals appeared quite similar to the initiatives of the Model Cities period and "as if we have learned nothing in ten years" (New York Times, December 18, 1977).

REFERENCES

ALLISON, G. T. (1971) Essence of Decision. Boston: Little, Brown.

BEAM, D. R. (1978) "Public Administration is alive and well—and living in the White House." Public Administration Review 38 (January-February): 72-77.

BLACK, J. T. (1978) Changing Economic Role of Central Cities, Washington, DC: Urban Land Institute.

CARTER, J. (1976) "Address on neighborhoods at Brooklyn College." New York, September 5.

Center for Responsive Technology (1974) Toward a Comprehensive Growth Policy Process. Washington, DC: Kettering Foundation.

COHEN, M. D., J. G. MARCH, and J. P. OLSEN (1972) "A garbage can model of organizational choice." Administrative Science Quarterly 17 (March): 1-25.

Congressional Quarterly (1978) Urban America: Policies and Problems. Washington, DC: Author (August).

DONOVAN, J. C. (1967) The Politics of Poverty. New York: Pegasus.

FRANKLIN, H. M. (1976) A Critique of the President's 1976 Report on National Growth and Development. Washington, DC: National Forum on Growth Policy.

HECLO, H. (1977) A Government of Strangers. Washington, DC: Brookings Institution.

JANIS, I. L. (1972) Victims of Groupthink. Boston: Houghton Mifflin.

MULLER, P. O. (1976) The Outer City: Geographic Consequences of the Urbanization of the Suburbs. Washington, DC: Association of American Geographers.

New York Times (1977a) "White House orders urban policy review; revises study panel." August 31.

——— (1977b) "U.S. report urges larger commitment to cities and the poor." November 7.

——— (1977c) "President delays urban aid rise." December 18.

President's Urban and Regional Policy Group (1977) Cities and People in Distress. Washington, DC: Author (November).

——— (1978) A New Partnership to Conserve America's Communities: A National Urban Policy. Washington, DC: Author (March).

SKIDMORE, OWINGS, and MERRILL (1977) Summary of Conservation Choices: Portland Energy Conservation Project. Washington, DC: U.S. Department of Housing and Urban Development.

SUNDQUIST, J. L. (1979) "Jimmy Carter as public administrator: an appraisal at mid-term." Public Administration Review 39 (January-February): 3-11.

THAYER, F. (1978) "The president's management reforms: theory x triumphant." Public Administration Review 38 (July-August): 309-314.

Urban Growth and New Community Development Act (1970) Statutes at Large. Vol 84.

U.S. Department of Housing and Urban Development (1977) New Directions for the President's 1978 Report on Urban Growth: Citizen Forums. Washington, DC: Author.

——— (1978a) A New Partnership to Conserve America's Communities: A Status Report on the President's Urban Policy. Washington, DC: Author (June).

——— (1978b) The President's National Urban Policy Report. Washington, DC: Author (August).

Wall Street Journal (1977) "The government's road show is a bomb." June 20.

Washington Post (1977) "Carter aides try to untangle urban policy." December 14.

——— (1978) "Carter signs 4 orders to provide administrative support to the cities." August 17.

——— (1979a) "Carter plans natural resources department." March 1.

——— (1979b) "Congressmen say urban employment pledge unfulfilled." March 11.

White House (1972) Report on National Growth. Washington, DC: Author (February).

——— (1974) Report on National Growth and Development. Washington, DC: Author (December).

——— (1976) 1976 Report on National Growth and Development. Washington, DC: Author (February).

White House Conference on Balanced Growth and Economic Development (1978) Final Report: Workshop Reports. Washington, DC: Author.

WILENSKY, H. L. (1967) Organizational Intelligence. New York: Basic Books.

7

The National Commission on Neighborhoods: The Politics of Urban Revitalization

JAMES T. BARRY

☐ UNDER PRESIDENT CARTER, there have been two attempts to arrive at overall policies for cities and toward revitalization issues. First, the Urban and Regional Policy Group was assembled early in 1977 to develop a comprehensive federal strategy for the cities. The chapter by Eric Stowe illustrates some of the difficulties involved in trying to formulate such a strategy. The National Commission on Neighborhoods was the second body given a mandate to examine revitalization approaches and design an overall policy. The present chapter explores selected aspects of the experiences of that body.[1]

To be more specific, the Neighborhood Commission was charged with responsibility for investigating the causes of neighborhood decline and recommending policies which would support revitalization. The commission was appointed by President Carter on December 19, 1977. Fifteen months later, on March 19, 1979, the commission submitted its final report to Carter and the Congress (National Commission on Neighborhoods, 1979). The report was submitted in three volumes. In Volume I, the heart of the report, the commission proposed over two hundred recommendations, with accompanying analytic justification, on how to preserve urban neighborhoods.[2]

Because revitalization is likely to become a more important policy issue, and because the National Commission on Neighborhoods, with the imprimaturs of the White House and Congress, attempted to establish a definitive policy toward revitalization, a history of the commission is worth retelling. The story will concentrate on substance: what the commission did and did not propose. However, an explanation for substantive decisions often is found in the process: the tangle of events through which people proceeded as they made their decisions. Among many influences which shaped the commission's work, the substance of the final report was most heavily influenced by those commissioners who were able to use the structure of the commission and the particular circumstances as means for realizing their own personal agendas for the commission. To paraphase the biblical aphorism, "Many interests were called, but few were fully represented."

PEOPLE BUILDING NEIGHBORHOODS: A SUMMARY ANALYSIS

Volume I of the commission's report contains an executive summary and six analytic chapters. In the executive summary, the commission presented more than a synopsis of its recommendations. It articulated its basic moral or political theme which can be characterized as left-of-center populism. As part of this, the commission defined the constituency for revitalization as the low- and moderate-income classes living in central cities. Explicitly rejected as appropriate policies were those which concentrated only on bringing the middle class back into the city, or on preparing marginal areas for reentry by mobile young professionals. Revitalization strategies must concentrate on increasing the ability of persons of low- and moderate income to improve their economic status and maintain their neighborhoods. No neighborhood, no matter how run down, should be written off.

The commission lamented the concentration of power in big government and big business, and demanded that power devolve to the neighborhood level. People, working through community and neighborhood organizations, should be empowered to control their own destinies. This would be accomplished by ending the unequal distribution of resources which favor the "haves" over the "have nots." Redistribution of resources including power would also be achieved by relying more on the initiative and self-help of people at the neighborhood level. This calls for developing appropriate economic structures at the local level and community organizations which are independent of outside influences.

At its first business meeting, the commission organized itself into five task forces which were to be responsible for specific substantive areas: (1) reinvestment, (2) economic development, (3) delivery of human services, (4) governance and citizen participation, and (5) legal, fiscal, and administrative obstacles to revitalization. Five of the six analytic chapters in the commission's Volume I were prepared as reports of these task forces. The remaining chapter, which analyzes relations between racial minorities and civil rights, on the one hand, and the neighborhood movement, on the other, was written by an ad hoc task force of minority commissioners.

The reinvestment task force focused on housing as the anchor for preserving and revitalizing neighborhoods. Explicitly disputing traditional views that low- and moderate-income families or racially changing neighborhoods presented unusual credit risks, the task force argued that in low- and moderate-income neighborhoods a substantial demand for home mortgage money was being artificially denied. Private lending institutions drew lines (referred to as "redlining") around neighborhoods within which mortgage loans would not be made. Redlining would be prohibited through regulatory controls which would prohibit geographical discrimination by the entire banking industry. Urban neighborhoods also have been put at a disadvantage by federal housing policies which favored the growth of the suburbs. These disadvantages would be righted by the imposition of targeting requirements on federal programs which would ensure that low- and moderate-income neighborhoods received adequate levels of public support. For example, the commission recommended that 75% of all of the Community Development Block Grant funds distributed by the Department of Housing and Urban Development (HUD) be allocated to low- and moderate-income neighborhoods.

The economic development task force concentrated its attention on neighborhoods which were economically distressed: where people were without jobs or in low-paying jobs; where commercial activity was nonexistent or traditional economic investments were not being made. That task force focused on justifying the case for alternative economic structures—community development corporations, local credit unions, community cooperatives, minority business enterprises—which would provide the investments, managerial resources, and entrepreneurial spirit necessary to revitalize depressed neighborhoods. The task force called for the reorganization and consolidation of federal programs which make economic development monies available. However, it declined to specify how this reorganization would be accomplished since, at the time its proposals were being made, President Carter's own cabinet reorganization plan for the Departments of Commerce and HUD was before Congress. For similar reasons, the task force declined to endorse or oppose the proposed National Development Bank, but did recommend that, if it were approved, there should be guidelines to insure that the bank made loans to low- and moderate-income neighborhoods. Finally, the task force made extensive recommendations which would target national economic planning and federal expenditures on low- and moderate-income neighborhoods in distressed cities.

The task force on human services made recommendations on the processes for delivering services rather than focusing on specific service delivery systems. Service delivery systems should be judged by standards of equity, accountability, and efficiency. The fragmentation of services must be overcome so that people receive all the services they need. In every neighborhood, regardless of its economic health, there are social networks of families, friends, and churches, which naturally provide care in times of need. The task force argued that professional providers of services should mesh them with those services already being provided by the natural support systems. The commission recommended that a demonstration project in the service delivery process be developed to test how professionals, neighborhood advocates, and service recipients could work together to determine service delivery arrangements. A model for this demonstration project would be the Neighborhood Housing Services program, which has successfully forged cooperative agreements between lenders and homeowners to make mortgage money available in moderate-income neighborhoods. The task force also recommended increases in funding for human services programs, and new allocation rules which would permit direct funding of neighborhood services programs.

These three task force reports were in basic agreement in their emphasis on recommendations supporting low- and moderate-income neighborhoods. The report of the task force on legal, fiscal, and administrative obstacles conceived of revitalization and its beneficiaries quite differently: a plethora of rules and regulations, obviously emanating from the public sector, tend to make revitalization too costly. The Income Tax Code favors new construction, and offers fewer incentives for rehabilitation. Building codes prohibit the use of less costly construction materials. The property tax system discourages home improvement.[3] The task force called for the elimination of all artificial constraints on individual initiative. The task force recommended that incentives be provided through the Internal Revenue Code to lure large investors into rehabilitation projects. The property tax system should be redesigned to shift more of the burden to the value of land and to develop an array of services which are paid for by fees charged to the users. Building codes would be more supportive of rehabilitation work if the codes relied upon establishing standards of performance which building materials must meet, rather than upon specifying the characteristics of the materials or if states and municipalities shifted to warranty programs which held the contractor liable for the soundness of the home and provided insurance coverage for the homeowner. The decline of the individual buildings or of whole neighborhoods could be arrested if the laws governing title and the procedures for recording titles were reformed to allow for a quicker turnover in ownership of neglected buildings.

The task force on governance and citizen participation operated almost by itself. Its chairman considered neighborhood organizations to be the heart and soul of the commission's deliberations. This task force argued that neighborhoods need organizations which are independent of outside influences and which have the capacity to work as equal partners with public and private bodies in all revitalization programs. Each neighborhood has its own mix of needs, aspirations, and resources. Therefore, no one type of revitalization program, no one form of neighborhood organizations are required to insure that the unique needs of the neighborhood are met. The report of this task force was a political blueprint for protecting neighborhoods from the power of larger political and economic interests. The task force was less interested in particular recommendations for economic development programs, reinvestment regulations, or property tax reform, and more concerned with a need to remind the commission that healthy, revitalized neighborhood always will require independent organizations.

To this end, the task force considered the funding of neighborhood organizations a priority commission issue. The task force went to great lengths to identify many sources of funding for neighborhood organizations, making detailed recommendations on how potential sources such as private corporate donations could be achieved, how the monopoly which the United Way exercises over solicitation at the workplace should be broken, and how churches could be approached for support. Disenthralled with the kind of citizen participation required by many federal program regulations, the task force proposed new requirements for citizen participation so that such groups would be consulted from the beginning of a planning process, thus not simply being used to ratify decisions already made, and preventing their use as a shield against more aggressive, independent neighborhood organizations.

Finally, the chapter on the neighborhood movement and urban minorities, entitled "Compounded Problems Facing Urban Racial Minorities," argued that minorities had a major claim upon revitalization programs. The neighborhood movement, it argued, must not become a vehicle for excluding minorities from white neighborhoods. Citizen participation regulations must explicitly insure that minorities are adequately involved. The disadvantages imposed upon minorities by racism require that assistance to minority neighborhoods be greatly increased.

There are subjects upon which the commission failed to provide analysis or recommendations. Urban education in gernal was not analyzed, nor was busing to achieve school desegregation mentioned. The report barely alludes to the problem of crime and fails altogether to propose any major anticrime programs. Throughout the report, the phrases "minority" or "racial minorities" appear. However, the report makes no effort to distinguish between the major racial minorities, blacks and Hispanics, or among the geographic and cultural differences within the Hispanic population. The conflicts which may exist between blacks and Hispanics, or among Hispanics, or between these minority groups and other new urban arrivals such as the Vietnamese, are not mentioned. The report, furthermore, makes no policy distinctions between the burdens which racism places upon lower-income as opposed to middle-income minority neighborhoods. The possibility that there is a large, quasi-permanent urban underclass, largely black, was not considered in any detail. And, finally, the report repeatedly makes assumptions about the extent of interracial cooperation which is both occurring and possible in central cities, claiming that there are major "convergent issues" which unite minority and white neighborhoods in a common purpose: this assumption was never seriously analyzed.

In the final report there are inconsistencies and contradictions which were allowed to remain. For example, the task force on obstacles made recommendations which would be of more immediate benefit to the affluent than to the low- to moderate-income classes. This certainly is true of the tax incentive proposals, which would only benefit the wealthy. The greater mobility which would be permitted by proposals to expedite the transfer of title would more clearly benefit real estate interests which are primed to take advantage of the market for gentrification than low-income residents of urban neighborhoods.

While the commission defined low- to moderate-income neighborhoods and families as their basic constituency, individual task forces were concerned about different ends of that spectrum. The reinvestment task force recommendations were of more relevance to those moderate-income neighborhoods in which there was some likelihood of a basic level of savings in local banks to support rehabilitation than would be true of some lower-income neighborhoods. At the same time, the economic development task force focused upon neighborhoods without income; that is, at the other end of the spectrum from reinvestment. However, there was no attempt to reconcile potential contradictions between these two task force reports: Should the antiredlining regulations proposed by the reinvestment task force apply equally to both moderate- and low-income neighborhoods? Should the economic stimulants suggested by the economic development task force be targeted exclusively upon low-income neighborhoods?

Among other gaps in the report, one can criticize the commission for not establishing priorities among the 231 recommendations it made. Is a recommendation about opening up work-place solicitation to groups other than the United Way as significant as a recommendation that 75% of all HUD Community Development Block Grant monies be allocated to low-

and moderate-income neighborhoods? Are recommendations for improving the delivery of services according to standards of equity more or less important than issues relating to the reorganization of economic development programs? How should a call for reorganizing the claims for revitalization funds of minority neighborhoods be weighed against proposals to revise building codes?

HOW THE REPORT WAS WRITTEN

On December 19, 1977, President Carter formally announced the appointment of twenty members to the National Commission on Neighborhoods. These members were:

Joseph Timilty, Chairman Massachusetts State Senator	David Lizarraga Community Leader, East Los Angeles
Dr. Ethel Allen Member, Philadelphia City Council	Vicki Mongiardo Community Organizer, Washington, DC
Anne Bartley Arkansas Commission on Historic Preservation	John McClaurghy Concord, VT
Representative James Blanchard D-Michigan	Arthur Naparstek Educator, Washington, DC
Nicholas Carbone President, Hartford City Council	Robert O'Brien Banker, Newark, NJ
Gale Cincotta Community Leader, Chicago	Representative Joel Pritchard R-Washington
Senator Jake Garn R-Utah	Senator William Proxmire D-Wisconsin
Harold Greenwood Banker, Minneapolis	Macler Shepard Community Leader, St. Louis
Maynard Jackson Mayor, Atlanta	Peter Ujvagi Community Leader, Toledo
Norman Krumholz City Official, Cleveland	Dr. Bathrus Williams Educator, Washington, DC

There was a balancing here of community leaders with local officials, of congressional people with bankers, of Republicans and Democrats. There was racial representation, with four blacks—Allen, Jackson, Shepard, and Williams—and one Hispanic—Lizarraga; there also was an ethnic balance with the Italians Carbone and Mongiardo, the Lithuanian-Greek Cincotta, and the Hungarian Ujvagi.

Of these twenty commissioners, eight played a major role in producing the final report. Timilty played a decisive role as chairman. The most important influence he exerted was through his leadership style. Experienced in the process of searching for winning majorities associated with legislative leadership, Timilty did not attempt to impose a report upon the commission but rather sought to manage the commission so that a majority emerged in the end. Therefore, the report was written so that those who exerted themselves got what they wanted into the report.

THE TASK FORCE CHAIRS

As a legislative leader, Timilty also knew the value of committees for accomplishing work and arranging for coalitions. Thus, early in the process of commission organization, he proposed the establishment of task forces, which Timilty then turned over to task force chairpersons. As time passed, the task forces became the crucibles out of which the work of the commission was forged. In some cases, the task forces became virtual baronies of their chairs. For example, Peter Ujvagi, chairman of the governance task force, joined the commission with a personal agenda which consisted solely of legitimating the process of neighborhood organizing. He exercised almost day-to-day control of his task force, operating as both chairman and staff director. Toward the end of the commission, Ujvagi often flew to Washington for all-night sessions with his staff, was "cutting and pasting" drafts of the report in the middle of the night, and had his principal writer in Toledo writing for several days.

The chairperson of the reinvestment task force, Gale Cincotta, exerted similar control. Cincotta was the driving force behind the antiredlining campaign. She had been instrumental in the passage of the Home Mortgage Disclosure Act, a major victory against redlining. Cincotta thought of reinvestment as her bailiwick. Her task force's report was written by a writer Cincotta chose. The writer was someone with whom she had worked before. Cincotta, that writer, and a key Cincotta aide controlled the writing of the task force report. No one in the commission, on the staff, or even on the reinvestment task force saw a draft of that report or heard a rumor of what would be recommended until the entire draft had been completed and then had been fully reviewed by Cincotta and her staff.

In the case of the obstacles task force, it was not the chairperson, the banker Robert O'Brien from New Jersey, but another Commissioner, John

McClaurghy, who dominated the task force. The legislation creating the commission directed the commission to investigate a number of issue areas, one of which was legal, fiscal, and administrative obstacles to revitalization, an idea which was borrowed from earlier papers written by McClaurghy. The subject of obstacles was McClaurghy's major interest on the commission. When the time to write the task force's report came due, McClaurghy retired to his home in Concord, Vermont, where he drafted the entire task force report. He tried with some diligence to involve the other members of the task force in writing the report but with little success.

The economic development report was written almost with a chairperson "in absentia." David Lizarraga was chairman of this task force. He was a logical choice, since he headed one of the largest community development corporations in the country—The East Los Angeles Community Union (TELACU). Lizarraga had been very active in the beginning, but, after a late May meeting in Los Angeles, Lizarraga withdrew from the commission work completely. However, before other interests absorbed him, Lizarraga and his task force had produced a detailed outline for the final report. In Lizarraga's absence, a staff writer worked from this outline. Drafts of the economic development report were available as early as September. They were circulated among the task force, but no written responses were ever received from any member of the task force.

The economic development task force was fortunate that it had an imaginative staff member to fill the breach of an absent chairperson. The human services task force was not so fortunate. This task force operated nominally under Dr. Ethel Allen, a black doctor from Philadelphia, and a member of the city council there. Allen had a long interest in health care delivery issues, and was a natural choice to chair the task force. Like Ujvagi and Cincotta, Allen exercised a strong control over her task force. However, over the summer and throughout the fall, Allen became deeply involved in Pennsylvania gubernatorial politics. The task force languished, and, just as with economic development, responsibility for the task force rested entirely with its staff member. This staff member had a December 10, 1978 deadline for the first draft of the task force's report. On the morning of that day, the staff person called the office to say there was no draft to submit. By reassigning a writer from another task force to human services, and by contracting with a Washington consulting firm in which Allen had great confidence, a task force report was written from scratch at the last minute.

THE BUDGET AND STAFF CRISES

This concentration of responsibility within the task forces was intensified by a financial and staff crisis which occurred in September of 1978.

In early September, the commission's chairman, Joseph Timilty, discovered that the commission faced a severe budgetary crisis. It had overspent in many categories, and, if the rate of expenditure was not curtailed, the commission might not have been able to set aside enough money to print and publish its final report. Timilty, with the Executive Committee of the commission, voted to cancel all further field hearings and conferences, to prohibit any new financial obligations for research, and to institute an immediate cut-back in the staff. Relations between the commissioners and staff had been strained for some time,[4] and the proposed reductions in the staff caused an immediate crisis. The staff produced its own plan in which it agreed to accept voluntary cuts in pay so that staff terminations could be forestalled for several months. However, the Executive Committee rejected that plan, insisting upon terminating some staff members immediately. In protest, most of the top staff members, including the Executive Director, resigned in protest.

The committee's legislative mandate had laid a heavy agenda upon the commission, specifying eight areas of research and a similar number of areas within which they were to make recommendations. The original drafts of the legislation would have appropriated two million dollars and granted two years to the commission. However, HUD Secretary, Patricia Roberts Harris, was opposed to the commission. While Harris did not succeed in scuttling the commission altogether, she was successful in getting both the money and the time cut in half. There was no commensurate reduction in the expected scope of work. However, the legislation recognized this problem, and provided for a possible six month and $500,000 extension for the commission.

The commission's application for an extension was being reviewed by Senator Proxmire's Banking Committee at the same time that the staff turmoil was occurring. The Banking Committee ultimately decided to grant only a three month extension with no additional funds. With the three month extension, the commission had five months remaining within which to complete its work. In mid-October, 1978, it had no staff writers, very little money, and with the exception of the economic development task force, not one word on paper toward a final report.

The commission did produce a report in those five months, but it was produced under severe disadvantages. Most important, the staff crisis left no alternative to relying upon each task force to complete its own work. With the exception of economic development, all of the task force reports were written between mid-October and the end of the year, usually in complete isolation from the commission as a whole. The commission had established a Policy Committee, cochaired by Commissioners Arthur Naparstek and Nicholas Carbone, and later a Report Drafting Committee under Naparstek, which were to have been responsible for coordination among the task forces. However, under the pressure of the financial and staff crises, both committees crumbled.

The commission's first opportunity to review drafts of task force recommendations and reports came on December 21 and 22. Few commissioners, other than those who had been involved with the writing of their respective task force reports, had much chance to review any of these drafts in advance. With the problems it faced, the human services task force only submitted a draft outline of a final report. Drafts of sections of John McClaurghy's obstacles report were circulated anywhere from one week to two weeks prior to the meeting. Reinvestment did not submit a report until the day before the meetings began, while the governance chairman and staff literally were "cutting and pasting" their draft together on the evening and early morning before.

Over these two days, the commission spent a total of twenty hours in meetings. During those twenty hours, the commissioners reviewed all the recommendations. Some recommendations were not approved. The draft report for obstacles called for the elimination of all rent control laws, which was politically unacceptable to a majority of the commissioners. The economic development draft proposed the consolidation of all neighborhood economic development programs under a new Assistant Secretary in the Department of Commerce. While consolidation was acceptable to them, the commissioners declined to specify where this new agency should be put since, at that time, President Carter and Congress were debating their own cabinet reorganization proposals.

For the most part, however, the recommendations were approved as drafted. The discussions proceeded almost as if they were negotiations between a task force chair and the rest of the commissioners. By this time, the task forces had become synonomous with their chairs. If he or she felt strongly about a recommendation, there was never a majority of the commissioners willing to vote against the wishes of the chair. Therefore, while the separate task force reports are logically consistent in their own ways, the commission lacked both a mechanism and the time to reconcile contradictions and inconsistencies among the chapters, or to make choices about which recommendations were more important than others. In a sense, the final report of the Neighborhood Commission was really seven separate reports. For, in addition to the five task force reports, the commission produced an executive summary and a chapter on urban minorities, both of which, like the task force reports, reflected the predominance of individual commissioners.

THE EXECUTIVE SUMMARY

The commissioners planned for the executive summary to be more than a simple synopsis of the task force reports. Instead, the summary was intended to be a document which could be published separately, in which the commission would advocate the case for neighborhoods in the most literary and declamatory fashion possible. Following the staff shakeup, the commission hired an editor who was responsible for drafting the executive summary. Between the December 21 meeting and the next scheduled meeting, January 19, 1979, the editor wrote a first draft of the summary. The draft was not available for review until the evening of January 18. It was unacceptable to several commissioners, but most importantly to Nicholas Carbone. His objection was to the manner in which the executive summary presented the theme of the commission.

The commission was driven by a desire to articulate a theme, a moral statement, a political principle, which conveyed about the commission's final report the same urgency as the most famous sentence in the Kerner Commission report: "Our nation is moving toward two societies, one black, one white—separate and unequal" (National Advisory Commission on Civil Disorders, 1967: 1). In the end, the commission settled for two themes, both of which were populist in tenor, but one more left-of-center than the other. On one theme, there was general agreement among the commissioners: within America's neighborhoods there was a vast resevoir of determination, self-reliance, and intelligence which the country needed to tap if the urban crisis was to be solved. President Carter's 1978 urban message expressed this theme when it said the federal government needed to draw upon "the sense of community and voluntary effort that I believe is alive in America, and on the loyalty that American feel for their neighborhoods" (White House, 1978).

However, there is in the executive summary a parallel theme in which the solutions to the problems of neighborhoods were to be found in redressing inequities which are endemic to the organization of economic and political power. The macroeconomic system and the political order exploit the poor, the disabled, the elderly, and racial minorities. Central cities are becoming containment areas in which society's unwanted are forced to live. Neighborhoods are misused by the forces which govern cities—state legislatures, municipal employees' unions, and the banks. Neighborhoods will be revitalized when these inequities are eliminated. This statement of a theme had far less support among the commissioners.

The commission first experienced conflict over these themes at its second meeting, held in Baltimore, the weekend of February 2, 1978. Among their items of business, the commissioners discussed the Urban and

Regional Policy Group (URPG), which Carter had earlier established to develop a national urban policy. The URPG had been circulating drafts of its report, which was scheduled for delivery to the President in March. The commission wished to influence the direction of the URPG policy, so it decided to seek a meeting with the cochairpersons of the URPG, HUD Secretary Patricia Harris and Stuart Eizenstat, who headed up the President's Domestic Council. For this meeting a policy document was to be written which would express the commission's views on a national urban policy.

A draft statement was written on the next day by a small group of commissioners led by Arthur Naparstek and Nicholas Carbone, to both of whom the theme of the Neighborhood Commission was most important. The document they drafted argued that the central problem of our age was the helpless encounter of the individual with big business and big government. People had become "isolated, dispensable, and alienated," not allowed to control their own lives. The draft argued that neighborhoods were being destroyed by inequities which are:

structured into the macro-economic institutions [and which provide] disincentives which often create negative preconditions making it difficult for working people, poor people, the elderly on fixed incomes, and the unemployed to maintain a standard of living [National Commission on Neighborhoods, 1978a: 24].

Examples were cited of disincentives, including redlining, utility companies which charge their highest per unit rates to their smallest customers, an inequitable income tax code, and the overreliance on property taxes.

As the discussion proceeded on this document, crucial divisions became evident. Joel Pritchard, Republican Congressman from Seattle, wanted labor included with business and government as creating problems of bigness. John McClaurghy, a Reagan Republican, objected to including tax inequities, arguing the subject was outside the purview of the commission. Norman Krumholz, a Cleveland city planner, criticized the document for excluding discussion of crime as a neighborhood problem. David Lizarraga, a Hispanic, insisted that in a discussion of residential exclusion and job discrimination, the term "minorities" should be used instead of "blacks." Harold Greenwood, a banker, disputed the phrase "neighborhood control of financial institutions" which appeared in the draft. Because of these and other objections, the commission decided against issuing a policy paper at that time.

The conflict between the two themes was not encountered again until almost a year later at a January 19, 1979 meeting, when the executive summary was reviewed. In June of 1978, however, the commission had issued an interim report in which the conflict was avoided in favor of stating the more generally acceptable of the two themes:

> A neighborhood is an open place where the human spirit can flourish because the scale is human; where people can feel that their environment is not beyond their control. A national neighborhood policy is required, not to impose yet another layer of Federal programs on skeptical communities, but to help create the preconditions in which local institutions can revive. This will require adjusting government policies and programs to make them more neighborhood-sensitive. It will require leaving much to people to decide for themselves [National Commission on Neighborhoods, 1978c: 1].

This almost mystical belief in neighborhoods, while generally acceptable, masked the conflicting views which emerged in Baltimore. However, this was the theme of the commission as it appeared in the first draft of the executive summary.

The draft ran into considerable opposition, principally from Gale Cincotta and Nicholas Carbone. At the January meeting, both commissioners criticized the summary for being too placid. Cincotta wanted the report to begin with a description of the oppression being experienced by low- and moderate-income neighborhoods. Carbone wanted the language of the summary to be more hard-hitting, accusatory, and angry.

This January meeting had been scheduled to be the commission's last. However, unable to resolve disagreements which arose at this meeting, Carbone's being among the most important, the commission decided to delay a final vote until February 8. Any commissioner wishing to propose changes in the report had to put them in writing and circulate them among the other commissioners before that meeting. For the February meeting, Carbone produced an alternative executive summary. Carbone had recently read a report of the Urban Bishops Coalition of the Episcopal Church, *To Hear and To Heed,* which painted an extremely bleak picture of the American city: "The unneeded, unvalued, and threatening minorities are to be isolated, ghettoized, and contained. . . . Economic, political and social decisions have been made to let the cities be the areas of confinement for this underclass." Carbone's alternative summary borrowed from this report, incorporating that analysis, and arguing further that the problems of neighborhoods were those of the inequities which derive from racial injustice and the political disdain with which the WASP upperclass historically regarded ethnic urban politics. Neighborhoods are declining because cities are not allowed to govern themselves. They have become the domestic colonies of a triumvirate of state legislatures, municipal unions, and banks which exploit the cities for their own gain.

The commission refused to simply substitute Carbone's draft for the one which had been presented to them earlier. Instead, they expanded the original draft to include Carbone's perspective. In this manner, the commission ended up proposing two themes: one a general and largely unobjectional, almost mystical populism; the other, a more aggressive, biting, partisan populism far harder for some to accept. In fact, the inclusion of this latter theme was in large part responsible for four commissioners refusing to sign the final report. Three Republicans, Senator Jake Garn, Representative Joel Pritchard, and John McClaurghy, and one Democrat, Representative James Blanchard, did not sign. While their published dissents object to the commission recommendations which call for increases in governmental regulations and federal spending, there was little doubt that they also refused to associate themselves with the populist rhetoric included at Carbone's insistence.

THE NEIGHBORHOOD COMMISSION AND CIVIL RIGHTS

On June 28, 1978, Benjamin L. Hooks, Executive Director of the National Association for the Advancement of Colored People, wrote to Commission chairman Timilty to express the NAACP's concern that:

In this period of media-propagated code words, we are concerned that "neighborhood" not signal "white" while "urban," "inner-city," or "ghetto" signal Black—thus effectively excluding the racial minorities from "neighborhood" programs altogether.

A few months earlier, at a summit conference in St. Louis, the NAACP had released a critique of President Carter's 1978 urban policy message, which Hooks reiterated in his letter:

We particularly oppose the use of the phrase "their communities" in that use of such language tends to reinforce the established beliefs of whites that in the communities in which they reside they have the right to exclude blacks and other racial minorities from occupancy.

Timilty responded to Hooks's letter on July 21, 1978. He wrote that "as an elected public official [I have] always sought to build coalitions across racial and ethnic lines." As to the position of the commission, Timilty wrote:

We view neighborhood-based issues such as red-lining, quality of public services, economic development and jobs, citizen participation, insurance availability, expanded housing choice, tax equity, better mass transit and community pride as issues that can unite constituencies of black and white working people. That is the great hopeful potential of the so-called neighborhood movement.

The Hooks/Timilty exchange characterizes two perspectives on civil rights over which the commission contested several times. Each time, Maynard Jackson, the first black mayor of Atlanta, Georgia, precipitated discussion of the issue.

Jackson first caused an engagement over race at the commission's meeting in Baltimore. A consultant had been hired to prepare a presentation on goals and options for the commission. The paper prepared by the consultant was a discussion of how any particular geographical area is conceived and reconceived in the minds of observers as the socioeconomic status of residents of the area might decline, revive, or remain stable. Thus, with declining areas, one talks about disinvestment or infiltration; in a reviving area, one talks about gentrification. In this paper, the ideal toward which to strive for a neighborhood was stability. The key to a stable neighborhood lay in its "coping mechanisms," such as "priests who keep young punks from becoming juvenile delinquents," or "neighborhood organizations that prevent the building of a freeway" (National Commission on Neighborhoods, 1978b).

Whether deserved or not, no presentation could have more directly inflamed the fears of minority commissioners that neighborhood stability meant racial exclusion. The consultant was not more than ten minutes into his oral presentation when Mayor Jackson erupted in anger. According to some present, Jackson labeled the presentation as blatantly racist. He excoriated the consultant, demanding that the commission have no further dealings with him. The commission had been definitely warned that the subject of racism would play a major part in their deliberations.

Jackson again raised the civil rights agenda in a letter to Timilty on July 12, 1978. In his letter, Jackson complained of the manner in which the commission was addressing "the issue of race and racism as it relates to urban neighborhoods." Maintaining that previous commission decisions had led him to believe that race and racism would "be incorporated into our final report," Jackson said, "I have seen very little evidence of this [the incorporation of race and racism] either in our field hearings or in the array of special issue conferences sponsored by the Commission." Jackson proposed that the commission convene a special issue conference on the "neighborhood problems of Afro-Americans and other minorities." Explaining the need for this conference, Jackson wrote:

> I firmly believe that if we fail to do this, the impact and credibility of our final report and recommendations to the President will be

viewed with great and insurmountable skepticism by urban Afro-Americans.[5]

Jackson's proposal did not meet with universal approval from among the black commissioners. Macler Shepard, President of Jeff-Vander-Lou, a not-for-profit community development corporation in St. Louis, and the only genuine neighborhood leader among the minority commissioners, in writing to Timilty objected strongly to convening this conference because "it would generate more heat than light and certainly provide no new information for those involved in crucial neighborhood and community development."

Despite Shepard's objections, the commission agreed to organize a conference on neighborhoods and race. However, the commission's financial crisis broke at the same time that planning for this conference began. The conference was canceled. At Jackson's insistence, nonetheless, a research contract was let with the Washington-based consulting firm, the Joint Center for Political Studies, to interview a wide range of civil rights leaders on the same subject. The research contract was expected to produce a report that would approximate the product anticipated from the canceled conference.

The consultant report, titled "Minorities in Neighborhood Development," was released just prior to the December 21 commission meeting. The report is an analysis of the views of national civil rights leaders on "neighborhood organizations," "the neighborhood movement," and the Neighborhood Commission. These "minority views" can be summarized in three propositions: (1) in every area of concern to the commission, such as reinvestment, economic development, or human services, racism places an additional burden of deprivation which must be borne by racial minorities; (2) white ethnic neighborhoods, by asserting their rights of recognition in the distribution of resources, are directly competing with minorities for an ever-shrinking pot; and (3) white ethnic neighborhood organizations, because they represent whites, more easily find allies among the upper echelons of the institutions which distribute these resources, thereby restricting the access of minorities.

The Joint Center report states:

A national commission on neighborhoods which advocates policy initiatives for neighborhood vitality is essential at a time when much of the nation's leadership is supporting the elimination or cutback of neighborhood programs. Such a commission must make minority concerns a top priority and *earn the trust of the minority if its existence is to make a difference* [emphasis added; Joint center for Political Studies, 1978].

The commission was being challenged to accept the proposition that urban minorities have a greater prior claim to the programs and policies which were being advocated in the name of neighborhoods. Unless the commission acknowledged this claim, minority leaders might oppose the final report. This threat had been raised in Jackson's letter of July 12, and was raised again in the Joint Center report.

The latter was considered at the commission's December meetings. Not much time passed at that meeting before Jackson again accused the commission of failing to adequately address the problems of race and racism. He demanded, in the form of motions, that the issues of race and racism be discussed in each chapter of the final report, as well as in a separate chapter on race which would be written. He also moved that a minority person be hired to write this new chapter. Finally, he moved that the commission officially accept the Joint Center report, which would be published in its entirety in an appendix. The commission voted to accept Jackson's motions, though some commissioners felt they were being blackmailed, while others saw the addendum as a prelude to a minority break with the commission.

The next scheduled meeting was January 19, 1979. Between the December and January meetings, each task force report was reviewed for its treatment of minority concerns. To fulfill Jackson's motion, a black writer, in fact a former staff member of the Joint Center and the author of a preliminary draft of the Center's report, was retained to write a "minority chapter" for the final report. The draft he produced consisted only of a reworking of the original Center report. The proposed chapter was circulated among the commissioners prior to the January meeting.

Soon after this meeting began, Maynard Jackson announced that he was considering refusing to sign the commission's final report,[6] confirming fears that some commissioners had had for some time. Jackson's dissent from the report, perhaps followed by all the other minority members, would have been a serious blow to the credibility of the commission and a major embarrassment to President Carter. Timilty called an immediate recess of the meeting, and confronted Jackson in the hall. Timilty told Jackson his support was vital and asked what was necessary to get it. Jackson replied that he needed more time to review the report. Timilty agreed to get him the time he wanted. Calling the meeting back to order, Timilty moved the commission into executive session, where, after a lengthy and often heated argument, the decision was made to postpone the final vote on the report until February 8.

After the full meeting adjourned, Jackson caucused with Cincotta, Naparstek, Ujyagi, and staff to determine what remained to be done on the minority issue. By this time, Jackson's major concern centered on his perceptions of the inadequacies of the minority chapter. This section would have to be strengthened before Jackson would feel comfortable in supporting the report.

During the ensuing two weeks, the minority chapter was substantially rewritten. Analysis and recommendations were drawn from the Joint Center report, the National Urban League's *State of Black America–1978* (Williams, 1979), and the response sent on March 22, 1978 by the Congressional Black Caucus to Carter's urban policy message.

The resulting chapter asserted that:

Families of all races pay a penalty for living in our central cities. However, of all the residents of distressed urban neighborhoods, blacks, Hispanics, and other racial minorities are the worst victims.

Later, the chapter claims:

The whites living in cities today either cannot move out or do not wish to move out. They will live in cities and in neighborhoods in which they will be sharing power with blacks, and both will either sink or swim together. To the whites in the central cities, racial exclusivity is counterproductive.

The chapter then documented, with data basically drawn from the Urban League report, the tremendous additional disadvantages endured by urban racial minorities. The chapter closed with recommendations taken, "in toto," from the Congressional Black Caucus, as well as from a small meeting of national civil rights leaders convened by the commission in late December of 1978. Some of the latter recommendations were:

- Assistance to minority neighborhoods through housing, economic aid, or other programs needs to be increased.
- Community and economic development programs shall mandate provisions for increased participation by minority and low-income community-based organizations.
- Program guidelines for citizen and neighborhood participation shall be clear and precise and specifically include minority and low-income neighborhoods.
- All levels of the public and private sector shall hire minority neighborhood development experts who have a commitment to and an understanding of the needs of their constituents.
- All national programs and policies which impact on minority communities shall be coordinated to prevent duplication and promote maximum community growth and development.

With remarkably little difficulty, Jackson accepted this draft of the minority report at the February 8 meeting. While maintaining that the chapter was still flawed, Jackson nonetheless agreed that the chapter represented a sincere effort to confront the problems of racial minorities. Without doubt, Jackson had prodded the commission a considerable distance away from its earliest convictions that the search for convergent issues was the most positive contribution to overcoming racism.

Just how far Jackson moved the commission becomes quite evident from reading the preface to the commission's report written by M. Carl Holman, Executive Director of the Urban Coalition and an acknowledged black leader. Holman notes that "neighborhoods have long been one way of defining class, race and national origin differences in America." He further points out that "in this time of economic strain and fiscal uncertainty, there is real danger of an exacerbation of racial, class and generational tensions." Nevertheless, about the growth of neighborhood pride and neighborhood organizations, Holman saw hope in "the stronger convergence of the people of working class, poor and 'mixed' or 'marginal' neighborhoods as they have found striking similarities in the problems they face."

Holman believed that there were problems which white, black, and Hispanic neighborhoods shared; Jackson forced the commission to highlight the dangers of racism. Holman saw hope in the possibility of multiracial coalitions; Jackson led the commission to the position that policies should focus specifically on protecting neighborhood organizations in minority communities. Holman saw economic distress as a bond that united people across class lines; the Jackson-commission position was that race was the priority issue. The commission's report might have more broadly represented the actual views of the white commissioners had they trusted their first instincts and sought to have their views on "convergent issues" more completely articulated but that did not happen.

CONCLUSION

Weather throughout the Midwest and across most of the East Coast allowed only a handful of commissioners, fewer than a quorum, to attend the February 8 meeting. The commissioners who were able to get to Washington could not take a final, binding vote on the report. As the "denouement," the staff mailed out to the commissioners a complete version of the report. They also received a ballot on which they could vote for or against the report, and have additional views on the report included if they wished. The ballots were to be returned by February 27. On March

1, the eleventh affirmative vote, the necessary majority, was received. Ultimately, sixteen members voted yes; four voted no.

The decision of January 19 to delay a final vote on the report was costly. The Government Printing Office had agreed that, if all manuscript were delivered by January 22, it would be able to have 2,000 printed copies of the commission's report ready by the March 19 deadline. Because the commission was unable to get the manuscript to the Government Printing Office on time, it was forced to submit a "xeroxed" copy of its final report to President Carter. The printed copies of the report did not actually become available until the middle of the summer of 1979 even though the report carries an April issuance date. Peter Ujvagi had made commitments to send out copies of the report to about 500 neighborhood organizations. With this exception, there was no systematic plan for distributing the report.

Originally, extensive media attention to the report and its release had been arranged for by the staff. The Chairman planned to send the staff around to key congressional offices and within the Administration to prepare the way for the report. Many neighborhood organizations around the country planned concurrent press conferences to announce the release of the report and their support for its recommendations. All of these plans had to be scrapped.

Thus, the official submission of the report was attended by the minimum of publicity. President Carter did direct Stuart Eizenstat to form an interagency task force to review the report. The task force was to prepare a package of legislative and administrative actions which the White House could take to be submitted by the middle of the summer.

However, the events of the summer of 1979 ran right over this task force. The Camp David domestic summit, the President's speech on the need for a rededication to American values and effective energy policy, and the shake-up of the cabinet left very little time in Washington for discussion of the Neighborhood Commission report. Three former commission members were among those invited to Camp David: Nicholas Carbone, Maynard Jackson, and David Lizarraga. However, there is little to suggest that the commission's work or their positions with respect to its recommendations figured in the conversations at Camp David.

While a majority voted for the final report, in actuality, most of the commissioners had not been heard from. Norman Krumholz, an appointed local official from Cleveland, had strongly held views about what the commission should recommend, but he was reticent in expounding them. David Lizarraga had some impact as a task force chairman but made no impact as a Hispanic on the overall direction of commission recommendations. Anne Bartley, with a deep interest in historic preservation, found few of her interests directly recognized in the final report.

Instead, the report was written by an activist core of commissioners who, drawing upon an aggressive urban populism, were speaking on behalf of the left wing of the Democratic party. In 1980, the country's mood appears to be politically much to the right of the report prepared by these commissioners. It is quite unlikely, therefore, that the report of the National Commission on Neighborhoods will have a serious impact on urban policy in the short run.

Despite its limitations, however, the report is an important document. If there exists a cadre of white neighborhood leaders who are committed to remaining in central cities, living in integrated neighborhoods, organizing around convergent, multiracial issues, open to sharing power with blacks and Hispanics, then there is much hope for the future of central cities. This contribution to our recognition of an important component of contemporary urban politics alone would justify the work of the commission.

NOTES

1. As a staff member of the Neighborhood Commission, the author had an opportunity to observe some of the events described here. For the most part, however, materials for this chapter come from the files of the commission, which are stored at the National Archives. The author did interview some individuals at length in connection with this study, but those interviews were not a primary source.

2. The second volume contains case studies of forty-five neighborhood organizations from across the country. Volume III presents an analysis of those case studies.

3. In the first draft of its report, this task force called for the elimination of rent control legislation as well as for the exemption of all rehabilitation from Davis-Bacon wage-rate regulation. The full commission voted these recommendations down.

4. Unfortunately, the commission had a considerable share of internal acrimony. In some cases those feelings persist, with the result that a number of persons approached refused to be interviewed. Because the author could not get "both sides of the story" on these matters, this chapter does not deal with some of the problems that marked the internal administration of the commission.

5. During its first six months, the commission traveled to many cities, including Baltimore, St. Louis, Chicago, Cleveland, Los Angeles, and Seattle, where it held hearings and toured selected neighborhoods. Individual task forces also sponsored small "issue conferences" such as a conference on multifamily housing sponsored by the reinvestment task force.

6. Representative James Blanchard, one of the commissioners who eventually decided not to sign the report, had informed the meeting of his intention earlier.

REFERENCES

Joint Center for Political Studies (1978) Minorities in Neighborhood Development: Report to the National Commission on Neighborhoods. Washington, DC: National Archives (December).

National Advisory Commission on Civil Disorders (1967) Report. Washington, DC: Author (July).

National Commission on Neighborhoods (1978a) "Draft policy statement." Washington, DC: National Archives (February 10).

——— (1978b) "The dynamics of neighborhood change: some buzzwords and operational definitions." Washington, DC: National Archives (February).

——— (1978c) "The case for neighborhoods: an interim report." Washington, DC: National Archives (June).

——— (1979) People, Building Neighborhoods: Report of the National Commission on Neighborhoods. Washington, DC: Author (April).

WILLIAMS, J. D. (1979) The State of Black America—1978. New York: National Urban League.

White House (1978) "Text of President Carter's announcement of urban policy." Washington, DC (March 27).

PART III

RECENT POLICIES AND PRACTICES

8

Distressed Cities:
Targeting HUD Programs

YONG HYO CHO and DAVID PURYEAR

☐ AS OTHER CHAPTERS IN THIS VOLUME INDICATE, the federal government has been playing a leading role in the effort to restore the economic, social, and fiscal stability of deteriorating cities by increasing the number and variety of grant-in-aid programs and the magnitude of resource commitments. The nature of urban decline and distress, however, make it clear that the federal government cannot do the job alone. Its efforts to support urban government finance and economic development is only a small part of the total set of influences acting on cities. Thus, the current (FY 79) federal budget for state and local assistance totals only $85 billion, or four-tenths of 1% of the nation's $2 trillion gross national product. Approximately $38 billion of this assistance is expected to go to local governments (direct aid and state pass-through combined) and perhaps $30 billion of this will end up in the budgets of local governments in metropolitan areas (Cho, 1978). Still less will reach central cities in those metropolitan areas.

President Carter's national urban policy clearly reflects the recognition of two essential realities associated with these budgetary facts. First, there is a limit to what the federal government can achieve alone. Therefore, it is necessary to forge an effective partnership with the private sector as well as with state and local governments. Second, to make maximum use of limited federal resources, it is necessary to target them where they are needed most (President's Urban and Regional Policy Group, 1978). This chapter focuses on the second of these two characteristics of a realistic urban aid program, the question of targeting, and examines the major programs of the Department of Housing and Urban Development in terms of how well they are targeted on urban distress.

There are two important aspects of such targeting. First, how well do the programs' eligibility criteria target on distress and second, how responsive are program allocations to differences in the *degree* of distress among the eligible participants? This chapter offers only suggestive evidence on the first of these issues and concentrates its attention on the second question.

The following section discusses alternative measures of urban distress and their relation to city characteristics. The next section examines five major HUD programs in terms of their targeting on distress, and the final section presents a brief summary.

MEASURING URBAN DISTRESS

Every urban area has a unique mix of social and economic characteristics which define its condition. Urban distress is one dimension of this condition, and measuring it is an imprecise science at best. The most common approach to measuring distress is to identify a set of basic characteristics which indicate distress in most cities and to create an index of these basic characteristics.

One of the most important indicators of city distress is loss of population. Loss of population is not a serious problem by itself, but two characteristics of city population loss make it an important indicator of distress. First, cities have lost a disproportionately large share of middle- and upper-income residents. Second, the loss of population is nearly always accompanied by a roughly proportional loss of employment, thus weakening the local economic base and local tax base.

In recent years, the nation's population growth has slowed. In contrast, differences in growth rates among regions, between central cities and suburbs within metropolitan areas, and between metropolitan areas and nonmetropolitan areas have increased. Growth in the South and West has outpaced that in the Northeast and the North Central regions. Within metropolitan areas, suburbs are generally growing, while central cities are declining. Large cities are growing more slowly than small cities and are less likely to be growing at all.

The population growth that has occurred in central cities in recent years is mostly the result of municipal annexation. For the nation as a whole, there would have been no growth in the population of central cities in the eighty-five largest SMSAs between 1960 and 1970 without the enlargement of municipal borders through annexation. Even in the South and West, where the population of central cities appears to be growing, most if not all of the reported growth in central city populations would not have occurred if 1960 municipal boundaries had been in effect in 1970 (Sacks, 1978).

Income level (per capita income) and the incidence of poverty (the proportion of population with incomes below the poverty level) also provide indications of urban health or distress. The income level is lower and the poverty incidence is higher among Southern cities than those in the West or North.

A third element of city distress is the level of unemployment. Large cities have significantly more severe problems with unemployment. In 1976, three-quarters of large cities had unemployment rates above 8% while just over 40% of medium and small cities had unemployment levels that high. Regional differences in city unemployment are equally drastic. In 1976, just over 70% of Northeastern cities and more than 50% of Western cities had unemployment in excess of 8%, while less than 30% of cities in the South and Midwest had more than 8% of their labor force unemployed.[1] The differences in unemployment rates between central city and suburb is also significant. The unemployment rate in suburban areas is generally lower than in their respective central cities regardless of the race, sex, or age characteristics of their populations.

A number of other characteristics are also related to various aspects of urban distress. They include the age of housing, crowded housing conditions, the proportion of dependent population, education levels, the local tax base, and the level of local government debt obligations. The President's 1978 National Urban Policy Report identified the following characteristics as typical of distressed cities:

(1) A very large number of the distressed cities have been manufacturing centers.
(2) The city is old in terms of the age of its residential and commercial construction.
(3) The city is land-locked by surrounding incorporated municipalities and cannot grow through annexation.
(4) The city is fully developed.
(5) The city contains a substantial minority population (U.S. Department of Housing and Urban Development, 1978: 40).

By using a combination of selected variables, including those discussed above, several indexes have been constructed to measure the conditions of city distress. For present purposes, only four of the available indexes will be reviewed and used for the analysis of HUD program targeting. The four measures are the Brookings Urban Condition Index, HUD's City Needs Index, the Urban Development Action Grant (UDAG) Distress Ranking, and the UDAG Impaction Ranking.[2] A description and comparison of these four indexes is presented below.

A SUMMARY OF FOUR DISTRESS INDEXES

The Brookings Urban Condition Index consists of three variables: the rate of population change, poverty, and pre-1939 housing (Nathan and Dommel, 1977). The larger the index value the greater the distress.

The City Needs Index, developed by researchers at HUD, is based on twenty variables representing community development needs, socioeconomic conditions associated with urban blight and substandard housing, and measures of economic and population loss. Through a factor analysis, these twenty variables are reduced to three factors: (1) age and decline, (2) density, and (3) poverty. The factor scores of these three factors are weighed and combined to create a single index (Bunce and Glickman, forthcoming). A larger index value is associated with greater distress in this index, too.

The UDAG Distress Ranking and UDAG Impaction Ranking were both developed as selection criteria for UDAG awards. The UDAG Distress Ranking is based on per capita income, job lag/decline, and unemployment, all three of which are weighed equally. A smaller index value indicates greater distress under this index.

The UDAG Impaction Ranking is also based on three variables: (1) the percentage of the total housing stock that was built prior to 1940, (2) the percentage of the total population that was in poverty in 1970, and (3) the degree to which the population growth rate lags behind that of all metropolitan cities. Each of these variables is weighed differently: .5 for the housing variable, .3 for the poverty variable, and .2 for the population growth variable (Jacobs, n.d.). As in the case of the UDAG Distress Ranking, a smaller index value indicates greater distress.

Each of these distress measures captures a slightly different combination of city distress conditions. High values on the Brookings and the City Needs Indexes and low values on the two UDAG indexes indicate the greatest distress. Grouped according to the population size of the cities, results are presented in Table 8.1. The largest cities, those with populations of 500,000 and over, are the most distressed by all four measures. For the other city size groups, the degree of distress and population size are not consistently related. The least distressed cities are not always the smallest cities. Cities with populations ranging from 50,000 to 100,000 are the least distressed according to the Brookings Index and the City Needs Index. In the case of UDAG Distress Ranking, the smallest cities are least distressed while the cities in the population range between 100,000 and 500,000 are least distressed according to the UDAG Impaction Ranking.

TABLE 8.1   City Distress Conditions by Population Size

City Population Size (000)	Brookings Index		City Needs Index		UDAG Distress Index		UDAG Impaction Ranking	
	N	Average Score	N	Average Score	N	Average Score	N	Average Score
500 and over	24	162.2	24	0.183	12	104.8	12	100.9
100–500	137	111.7	134	−0.110	39	145.2	39	133.6
50–100	246	84.0	202	−0.338	26	139.6	26	102.9
Below 50	118	119.3	114	−0.132	13	169.1	13	114.1

Source: HUD data sources and data provided by the Brookings Institution.

A regional pattern of city distress also exists. As expected, cities in the Northeast are most distressed by all measures, as shown in Table 8.2. Western cities are least distressed by all measures except the UDAG Distress Ranking which indicates less distress in Southern cities than in Western cities. Cities in the North Central Region are the second most distressed according to all except the City Needs Index which shows Southern cities as the second most distressed group.

TABLE 8.2   Regional Pattern of City Distress Conditions

Regions	Brookings Index		City Needs Index		UDAG Distress Index		UDAG Impaction Ranking	
	N	Average Score	N	Average Score	N	Average Score	N	Average Score
Northeast	92	201.75	96	+.052	29	82	29	59
New England	42	151.24	45	−1.86	8	69	8	78
Middle Atlantic	50	244.18	51	+.262	21	87	21	52
North Central	122	113.70	129	−.381	25	165	25	120
East North Central	92	116.67	97	−.340	20	158	20	121
West North Central	30	104.57	32	−.505	5	196	5	120
South	141	100.50	145	+.017	21	191	21	154
South Atlantic	62	113.54	65	−.021	9	201	9	121
East South Central	23	110.35	24	+.089	4	183	4	151
West South Central	56	82.02	56	+.029	8	182	8	193
West	106	40.65	108	−.619	10	168	20	203
Mountain	22	41.97	22	−.716	2	236	2	186
Pacific	84	40.31	86	−.594	8	151	8	207

Source: Same as Table 8.1.

Finally, the four distress measures are intercorrelated. The Brookings Urban Condition Index and UDAG Impaction Ranking are most closely correlated ($r = .893$) while the next closest relationship is found between the Brookings Index and City Needs Index ($r = .7484$). The relationship is weakest between the City Needs Index and UDAG Distress Ranking ($r = .4246$). Nevertheless, the correlations are all highly significant statistically. These correlations confirm that the four indexes overlap to a considerable extent but in varying degrees.

HUD PROGRAMS AND URBAN DISTRESS

Five HUD programs have been selected for an assessment of the degree to which they target resources on the distress conditions of the cities. They are the Community Development Block Grant Program (CDBG), Urban Development Action Grant Program (UDAG), the Low Income Housing Assistance Payments program (Section 8), the Low-Rent Public Housing Program, and the Government National Mortgage Association (GMNA) Targeted Tandem program.

The CDBG and UDAG programs are oriented largely toward a broad range of community and economic development efforts. The others are devoted to housing development and housing assistance, particularly for low-income and elderly households. Since the objectives and characteristics of these programs are described in detail in this volume and elsewhere (Bunce and Glickman, forthcoming; Nathan et al., 1977; Jacobs, n.d.), the nature of these programs will be discussed only briefly here.

The distribution pattern of the five HUD programs will be examined in two ways. First, the pattern of distribution among central cities, suburbs, and nonmetropolitan areas, as well as among the four census regions will be examined. Then, the distribution pattern will be examined in relation to the degree of distress.

CDBG AND UDAG: GEOGRAPHICAL ALLOCATIONS

The geographical distibution pattern of CDBG funds reported here is based on allocation data for FY 77 and FY 78. The total funds available for allocation were $3,148 million for 1977 and $3,500 million for 1978. A total of 559 metropolitan cities and urban counties with populations in excess of 200,000 received CDBG entitlements based on the allocation formula. The remainder of the CDBG funds were allocated on a discretionary basis to the SMSA balance and to nonmetropolitan areas. Both these categories received set-aside amounts. The resulting allocations are summarized in Table 8.3.

TABLE 8.3 Distribution of CDBG Allocations by Type of Recipient and by Region

	Per Capita $ Amounts			Percentage Shares		
	(1) 1977 Allocation (1974 Formula and Hold Harmless)[a]	(2) 1978 Allocation (Dual Formula and Hold Harmless)[b]	(3) Dual Formula (Projected 1980)[c]	(4) 1977 Allocation (1974 Formula and Hold Harmless)[a]	(5) 1978 Allocation (Dual Formula and Hold Harmless)[b]	(6) Dual Formula (Projected 1980)[c]
SMSA	16.11	17.87	19.10	81.3	82.5	81.3
Metropolitan cities (559)	23.82	26.56	28.36	61.9	63.5	62.5
City type						
Central cities (381)	25.79	28.59	30.48	55.9	56.5	55.5
Noncentral cities over 50,000 population (178)	14.39	16.64	18.22	6.0	7.0	7.0
Region[d]						
Northeast	28.39	32.42	34.22	18.8	20.0	19.0
North Central	21.77	28.51	31.17	13.9	16.8	16.9
South	25.71	24.73	25.57	17.5	15.5	14.7
West	18.18	18.29	20.06	10.9	9.5	9.4
Remainder of SMSA	7.94	8.54	9.12	19.4	18.9	18.8
Urban counties (entitled)[e]	9.94	11.27	13.37	10.6	12.5	12.0
SMSA balance (discretionary)	6.39	6.64	5.81	8.8	8.0	6.8
Non-SMSA (discretionary)	9.70	10.17	11.74	18.7	17.5	18.7[f]
U.S.	14.34	15.75	17.57	100.0	100.0	100.0

a. The $3.148 billion appropriation in FY 77 was distributed on the basis of the 1974 single formula and hold harmless averages.
b. The $3.5 billion appropriation in FY 78 was distributed on the basis of the dual formula and hold harmless averages. Hold harmless credit in FY 78 was equal to two-thirds of the excess of the hold harmless amount over the basic dual formula amount.
c. These are full-formula (i.e., no hold harmless) amounts based on a projected 1980 appropriation of +4.8 billion.
d. Because 8 cities in Puerto Rico are excluded, the regional percentages will not sum to the total percentage for all metropolitan cities.
e. Data not available for breakdown of hold harmless between urban counties and the SMSA balance.
f. The non-SMSA account falls below 20% because the SMSA balance account includes a minimum set aside which is not divided on a 80-20 basis upon the SMSA and non-SMSA account.

In both of the years under examination, central cities received more per capita than noncentral cities, urban counties, or nonmetropolitan areas. Among the metropolitan cities, the Northeast region received more than the other regions in both years. The South received the second highest per capita amount in 1977 while the North Central received the second highest amount in 1978. The allocations of CDBG funds to metropolitan central cities and to the Northeast and North Central regions were more favorable in 1978 under the dual formula introduced that year than in 1977.

UDAG is a discretionary grant awarded to eligible cities upon the approval of an application submitted by an eligible city. A total of 319 cities are eligible for UDAG and through October 1978, action grants were approved for 101 cities. The biggest recipients were central cities with 75.2% of the total of $488 million granted up to that time, with suburbs and nonmetropolitan areas sharing the remainder equally (12.4% each). Because UDAG is intentionally targeted on distressed cities, it is not surprising that the largest share of the grants went to the Northeast, nearly 42%, while the North Central Region was second with about 30%. The South and West received slightly more than 14% each. (For greater detail on allocations, see the chapter by John Gist in this volume.)

CDBG AND UDAG TARGETING ON URBAN DISTRESS

The distribution pattern of the CDBG and UDAG funds among the regions of the nation and among central cities, suburbs, and nonmetropolitan areas generally indicates that a greater share of grant funds goes to the more distressed regions and the more distressed areas within those regions.

CDBG funding for 1978 and CDBG entitlement for 1980 as well as UDAG funding through October 1978 are grouped and averaged for each quintile of the four distress indexes. The results appear in Table 8.4. The 1978 CDBG allocation and the 1980 CDBG entitlement are systematically targeted to the degree of urban distress according to all of the indexes except the UDAG Distress Ranking. For example, the per capita CDBG allocation in 1978 for the most distressed quintile of the Brookings Urban Condition Index was $45.27 compared with $28.22 for the second quintile, $23.56 for the third quintile, $17.03 for the fourth quintile, and $11.86 for the least distressed quintile. Although the interquintile difference is not as sharp, the City Needs Index and UDAG Impaction Ranking show a similarly consistent pattern of targeting in the distribution of CDBG funds.

The distribution of UDAG funding on a per capita basis is not consistently related to any of the distress indexes, including the UDAG Distress

TABLE 8.4 City Distress Conditions and Distribution Pattern of Urban Development Program Funds

Urban Condition Index by Quintiles	CDBG Funding and Entitlement				UDAG Funding up to 3rd Round		
	N	Average Index Score	1978 Allocation ($)	1980 Entitlement ($)	N	Average Index Score	Funding Approval through 1978 ($)
Brookings Urban Condition Index							
1	105	274.2	45.27	44.30	18	340.19	44.44
2	105	130.5	28.22	28.24	18	254.75	18.03
3	105	72.2	23.56	22.66	18	185.13	33.01
4	105	31.2	17.03	17.04	18	137.99	21.46
5	105	5.6	11.86	12.81	18	82.89	21.89
Total	525	102.7			90	200.19	27.77
City Needs Index							
1	96	.697	39.99	40.30	17	.846	25.53
2	94	.109	35.07	32.67	17	.469	28.49
3	95	-.225	25.30	25.44	18	.218	40.79
4	94	-.519	16.97	18.15	17	.021	17.39
5	94	-1.066	13.12	13.48	21	-.0270	26.22
Total	474	-.198	26.32	26.24	86	.256	27.31
UDAG Impaction Ranking							
1	18	21.4	50.85	50.99	18	21	29.96
2	18	62.0	50.71	42.99	18	62	39.25
3	18	106.2	39.28	35.47	18	106	22.39
4	18	155.4	28.16	29.36	18	155	26.09
5	18	242.7	24.90	23.37	18	243	21.14
Total	90	117.5	38.78	36.43	90	118	27.77
UDAG Distress Ranking							
1	18	30.0	46.45	46.53	18	30	22.41
2	18	87.9	50.22	40.33	18	88	22.20
3	18	139.1	30.13	31.87	18	139	29.78
4	18	193.7	36.93	35.01	18	194	23.38
5	18	257.6	30.18	28.43	18	258	41.06
Total	90	141.7	38.78	36.43	90	142	27.77

Source: HUD data sources.

Ranking and UDAG Impaction Ranking themselves. The cities in the most distressed quintile of the Brookings Urban Condition Index received the highest per capita UDAG funding, but the cities in the second most distressed quintile received the lowest per capita funding among the five groups. In any case, it is essential to remember that grant dollars per capita is not a particularly appropriate way to assess the degree of "targetedness" of any project grant program. Because UDAG funding is based on specific actionable projects proposed by the eligible cities and the funding magnitude for individual projects almost always increases less than proportionately with city population size, large cities and hence the more distressed cities receive a smaller share than they would likely receive from an entitlement formula based on distress. This is inherent in the nature of UDAG and should not be interpreted to indicate anything more than the need for complementary formula programs (such as CDBG). The basic rationale for UDAG is a quick response to project opportunities in distressed areas and a formula grant would destroy this flexibility. Finally, it should be noted that the UDAG eligibility formula is more highly targeted on distressed cities than other federal government grant programs, so all the awards go to cities with a significant degree of distress.

This point is clearly corroborated by Gist's study of UDAG. Gist's findings show that more cities in the higher distress quintiles tend to receive Action Grants (Gist, Table 10.2). For example, fifty-one cities in the top distress quintile have received Action Grants, while only fourteen cities in the fifth distress quintile have received Action Grants.

LOW-RENT PUBLIC HOUSING: GEOGRAPHIC ALLOCATION

The low-rent public housing program is one of the largest and oldest assisted housing programs. This program was established by the U.S. Housing Act of 1937 to provide "decent, safe, and sanitary" housing for low-income families and individuals. By 1977, there were nearly 1.1 million conventional public housing units and some 17,000 units of leased housing. Throughout the history of this program, over 9,700 projects have been built in 4,200 localities by 2,810 public housing agencies, and nearly four million people are presently housed.

More than 70% of conventional public housing units are located in the Northeast and South while only 8.4% are located in the West. In contrast, the Section 23 leased housing program is most popular in the West where more than 37% of the leased housing units are located.[3]

As might be expected, nearly two-thirds (64.6%) of total public housing units are located in central cities while only 13.1% are located in suburbs. The remaining 22.3% are located in nonmetropolitan areas.

TABLE 8.5   Total Outlays for Public Housing, FY 77 by Region and Metropolitan/Nonmetropolitan Area (dollars in millions)

			Program Components		
Location	Percentage of Total	Total	Conventional Debt Service Payments	Conventional Operating Subsidies	Leased Housing
Region					
Northeast	41.5	660.0	341.7	270.0	48.3
North Central	19.9	317.0	189.0	103.7	24.2
South	26.3	418.0	263.0	106.9	48.2
West	12.3	195.9	84.0	40.7	71.2
Total	100.0	1590.9	877.7	521.3	191.9
Metro/Nonmetro					
Central city	67.6	1075.2	567.0	384.2	124.0
Suburbs	13.7	218.3	115.0	78.2	25.1
Nonmetro	18.7	297.4	195.7	58.9	42.8
Total	100.9	1590.9	877.7	521.3	191.9

Source: HUD Office of Finances and Accounts and HUD Offices of Assisted Housing.

In dollar terms, HUD outlays totaled nearly $1.6 billion for FY 77 including $877.7 million for conventional debt service payments, $521.3 million for conventional operating subsidies, and $191.9 million for leased housing.

LOW-RENT PUBLIC HOUSING:
TARGETING ON URBAN DISTRESS

Public housing units and operating subsidies have been grouped and averaged according to the quintiles of the four distress measures. The results appear in Table 8.6. In order to make the data comparable, the operating subsidies are presented in per capita terms and public housing units are computed as a percentage of population.[4]

Per capita operating subsidies and public housing units as a percentage of the population are remarkably consistent with the distress quintiles of the Brookings Urban Condition Index and the City Needs Index. Per capita operating subsidies, for example, ranged from $18.21 for the most distressed quintile of the Brookings Urban Condition Index to only $4.57 for the least distressed quintile of the same measure. These public housing variables are only slightly less closely related to the distress quintiles of the UDAG Distress and Impaction Rankings.

TABLE 8.6 City Distress Conditions and Distribution Pattern of Housing Programs: Public Housing

Urban Distress Index by Quintiles	N	Average Index Score	Public Housing Units/City	Operating Subsidies/City ($000)	Operating Subsidies per Unit ($)	Operating Subsidies per Capita	PH Units as a Percentage of Population
Brooking Urban Condition Index							
1	23	385.5	4,382	3,660.2	725	18.21	2.16
2	22	245.0	11,253	10,745.1	863	12.51	1.31
3	23	153.1	2,984	1,820.1	533	8.88	1.47
4	23	94.3	3,098	1,565.0	491	6.12	.84
5	22	41.1	2,773	1,462.2	515	4.57	.75
Total	113	184.5	4,883	3,810.6	624	10.09	1.22
City Needs Index							
1	23	.975	4,883.3	4,195.0	745	17.95	1.83
2	23	.491	10,893.6	9,892.8	731	12.24	1.40
3	22	.235	2,886.1	1,612.2	531	7.93	1.27
4	23	-.029	2,818.4	1,641.3	553	6.79	.81
5	23	-.398	2,578.3	1,485.0	558	5.22	.71
Total	114	.255	4,828.8	3,784.2	624	10.09	1.21
UDAG Impaction Ranking							
1	10	17.5	3,880	3,527.2	759	15.16	1.78
2	9	54.8	5,696	5,059.1	768	14.98	1.97
3	10	85.0	17,939	15,308.1	723	10.52	1.28
4	10	134.5	4,927	4,376.4	722	11.50	1.41
5	10	235.9	4,349	2,129.1	519	5.81	.72
Total	49	106.6	7,392	6,100.8	697	11.52	1.28
UDAG Distress Ranking							
1	10	20.4	5,964	6,117.9	912	18.74	1.55
2	9	65.2	16,693	14,554.7	811	10.88	1.26
3	10	116.8	7,213	5,119.2	579	10.18	1.19
4	10	181.8	3,966	2,047.7	506	6.44	1.08
5	10	256.8	4,054	3,510.0	688	11.32	1.41
Total	49	129.5	7,392	6,100.8	697	11.52	1.28

Source: HUD data sources.

Table 8.6 clearly suggests that more public housing units and more operating subsidies are provided for the more distressed cities. However, the implication of this particular finding is not quite clear. In view of the fact that public housing units have been cumulatively developed over the past forty years, the extent of a public housing program cannot be construed as reflecting only the current needs of distressed cities.

SECTION 8: GEOGRAPHIC ALLOCATION

Section 8 is a new program, established by the Housing and Community Development Act of 1974 to provide new construction, substantial rehabilitation, or existing housing for low-income and elderly families and individuals. The allocation of Section 8 program units is governed by a complex set of legislative requirements, most importantly by Section 213 of the Housing and Community Development Act of 1974. First, 20-25% of the funds available must be allocated to nonmetropolitan areas. Second, the types of units allocated, namely, existing, new construction, or substantial rehabilitation, must conform as closely as possible to the specifications of a local Housing Assistance Plan (HAP). The funds must be allocated according to a formula based on such housing need factors as population, poverty, housing overcrowding, housing vacancies, substandard housing and "other objective measurable conditions." These variables are the primary foundation that determines the "fair share" allocation factor for each jurisdiction along with construction cost differences. In addition, statutory and administrative "set asides" affect the allocation of Section 8 funds.

A total of 1,235,600 units of Section 8 housing reservations had been made cumulatively through FY 78. This number exceeds the units of low-rent public housing currently under management (1,216,000 units). The distribution pattern of Section 8 reservations is quite different from the distribution pattern of low-rent public housing. Section 8 reservations for new construction are rather evenly distributed among the four census regions as well as among central cities, suburbs, and nonmetropolitan areas (see Table 8.7). Existing Section 8 reservations are also evenly distributed among the regions. Only substantial rehabilitation reservations are concentrated in the Northeast and in central cities.

An examination of the distribution of Section 8 new construction and substantial rehabilitation contract amounts (not shown here) indicates that about three-quarters of the contract amounts are allocated to the Northeast and North Central regions. The share for central cities and suburbs is slightly greater than three-quarters. In fact, the suburban share of Section 8 programs is relatively greater than the other major HUD programs.

TABLE 8.7   Units of Section 8 Reservations by Metropolitan Location and Region (cumulative through FY 78)

	New Construction	Substantial Rehabilitation	Existing
Housing Units (in 000)			
Total	480.1	75.5	681.0
Location not coded	(.9)	(.1)	(.9)
Allocated	479.2	75.4	680.1
Region (%)			
Northeast	32.2	55.6	28.2
North Central	33.3	18.4	20.4
South	21.3	15.7	27.8
West	13.4	10.4	23.7
Metropolitan Location (%)			
Central Cities	36.2	62.7	45.8
Suburbs	33.7	27.3	34.8
Nonmetropolitan	30.1	10.0	19.5
Total	100.0	100.0	100.0

Source: Adapted from HUD data sources.

SECTION 8: TARGETING ON URBAN DISTRESS

The allocation pattern of Section 8 programs corresponds to the degree of city distress conditions in a moderate way. As shown in Table 8.8, all four measures of Section 8 distribution—units reserved per city, total reserved amounts per city, reserved units per 1,000 existing housing units, and reserved amounts per capita—tend to be greater for the upper distress quintiles of the Brookings Urban Condition Index and the City Needs Index. However, the targeting is not apparent for UDAG Impaction and UDAG Distress Rankings.

GOVERNMENT NATIONAL MORTGAGE ASSOCIATION (GNMA) TARGETED TANDEM: GEOGRAPHIC ALLOCATION

The Government National Mortgage Association (GNMA) is the federal agency that assists the secondary market for home mortgages by supplying liquidity for mortgage investments. GNMA buys home mortgages such as assisted and unassisted mortgages under the Tandem Plan, below-market interest rate mortgages, and conventional mortgages. GNMA initiated the Tandem Plan Special Assistance functions in 1966 to provide mortgage

TABLE 8.8 City Distress Conditions and Distribution Pattern of Housing Programs: Section 8

Urban Distress Index by Quintiles	N	Average Index Score	Section 8 Units rsvd Per City	Total Reserve Amount ($ 000)	Section 8 rsvd Units per 1000 Existing Units	Reserve Amount per Capita ($)
Brookings Urban Conditions Index						
1	51	326.3	2,382	7,036	24.5	25.63
2	50	170.2	2,012	6,282	24.2	25.08
3	50	102.4	1,588	3,722	18.8	15.42
4	50	63.7	1,157	2,664	17.7	15.26
5	51	24.9	1,156	2,770	18.5	12.89
Total	252	137.8	1,660	4,498	20.7	18.86
City Needs Index						
1	49	.764	2,877	9,427	24.5	23.05
2	48	.218	1,290	3,268	24.1	24.86
3	49	-.066	2,224	5,252	20.4	18.20
4	48	-.325	1,119	2,577	18.4	14.82
5	48	-.697	1,032	2,615	18.0	13.90
Total	242	-.018	1,715	4,650	21.1	18.98
UDAG Impaction Ranking						
1	15	21.8	2,413	7,548	24.2	28.58
2	16	70.6	3,931	11,419	25.4	26.83
3	15	113.8	3,233	12,416	19.7	20.61
4	15	160.7	2,294	5,411	23.7	21.02
5	16	243.8	3,855	8,560	18.8	18.25
Total	77	123.1	3,164	9,095	22.3	23.04
UDAG Distress Ranking						
1	15	32.9	3,522	10,748	21.8	25.87
2	15	92.5	3,521	13,757	20.5	20.55
3	16	143.0	4,593	10,671	27.2	25.93
4	14	190.5	1,962	4,390	19.3	15.59
5	17	258.5	2,180	5,913	22.4	26.18
Total	77	145.9	3,164	9,095	22.3	23.04

Source: HUD data sources.

credit in capital-short markets and to help limit the effects of cyclical declines in mortgage credit availability and residential construction. The Tandem Plan pairs government investors such as the Federal National Mortgage Association (FNMA) and the Federal Home Loan Mortgage Corporation (FHLMC) with the special assistance functions of GNMA. The Tandem Plan has been used to support specific FHA-insured and VA-guaranteed housing programs as well as to provide mortgage credit at below-market rates during periods of tight credit.

Targeted Tandem was initiated in FY 78 with the objective of attracting or holding middle-income renters in distressed cities, that is, cities eligible for Urban Development Action Grants. An annual program allocation of $500 million was made for FY 78 and FY 79, and $484.4 million was actually committed in FY 78, for 17,698 units in 115 projects in 62 cities.

The Targeted Tandem allocation is more heavily concentrated in the Northeast and South among the regions and is predominantly central city oriented. Nearly 90% of housing units and program expenditures go to central cities.

GNMA TARGETED TANDEM: TARGETING ON URBAN DISTRESS

Although GNMA Targeted Tandem is heavily central city oriented, the allocations of housing units and program fund commitments are not related to any of the four distress measures (see Table 8.9). In interpreting this result, it is helpful to understand the precise nature of the program allocation procedure for GNMA Targeted Tandem. It is quite different from the other four HUD programs discussed above because of the demand-oriented nature of the program. The final allocation is a hybrid of the eligibility criteria (location in a UDAG distressed city) and the number of individuals in the various eligible locations who apply for the program. The only part of the allocation process under HUD control is the eligibility criteria. As already noted, the program achieves a substantial degree of targeting on central cities and on the Northeast. These concentrations are suggestive of a high degree of targeting on distress. The allocation of funds among eligible cities, however, depends on individual applications, not on HUD. It is determined by public demand for the program in the form of mortgage applications. At this level, as Table 8.9 indicates, the most distressed cities are not generating as much demand as some less distressed cities, according to all four indexes of distress. Given the nature of the GNMA Targeted Tandem allocation process, this is much less surprising than it would be for any of the other HUD programs discussed here, all four of which exhibited some significant degree of targeting on urban distress.

TABLE 8.9 City Distress Conditions and Distribution Pattern of Housing Programs: GNMA

City Distress Condition Index by Quintiles	N	Average Index Score	Number of Housing Units per City	Total Funds Commitment per City ($ 000)	Fund Commitment per Capita ($)	Fund Commitment per Unit ($)
Brookings City Condition Index						
1	10	277.8	351	8,720	14.32	24,000
2	10	193.6	492	17,400	13.70	26,000
3	10	134.7	229	5,880	36.87	27,000
4	9	92.4	191	4,256	23.54	24,000
5	10	55.0	422	10,480	5.67	26,000
Total	49	151.9	340	9,451	19.41	25,000
City Needs Index						
1	10	.908	362	9,390	19.86	24,000
2	10	.514	521	17,510	12.60	27,000
3	10	.297	212	4,310	29.80	22,000
4	9	-.041	325	8,789	35.39	26,000
5	10	-.332	270	6,850	38.60	27,000
Total	49	.290	338	9,382	18.56	25,000
UDAG Impaction Ranking						
1	4	55.5	410	12,075	28.98	28,000
2	4	74.5	644	17,900	12.87	24,000
3	4	119.3	750	28,625	12.46	31,000
4	4	170.5	130	3,525	13.31	25,000
5	4	264.5	322	8,075	37.74	24,000
Total	20	136.9	451	14,040	15.32	27,000
UDAG Distress Ranking						
1	4	47.0	741	28,750	11.75	33,000
2	4	102.3	528	16,400	17.58	25,000
3	4	139.8	319	7,275	14.27	24,000
4	4	197.8	236	5,900	13.26	26,000
5	4	246.3	432	11,875	47.56	24,000
Total	20	146.6	451	14,040	15.32	27,000

Source: HUD data sources.

207

SUMMARY AND CONCLUSIONS

How well are HUD programs targeted to the distressed conditions of our cities? Is the allocation of HUD programs proportionate to the degree of city distress? Answers to these questions are dependent upon two things: (1) the particular program at issue, and (2) the particular distress measures considered.

In terms of the regional patterns of program allocation, the Northeast region is almost always the largest recipient of program shares, followed by the North Central, South, and West. The largest or dominant share of these programs also goes to central cities. Usually nonmetropolitan areas and suburbs are a distant second and third in the program shares they receive. This pattern is appropriate for two reasons. First, the target populations for these programs are concentrated in central cities and in the Northeast. Second, the federal budget as a whole tilts against central cities and it is appropriate for the lead urban agency, however small its programs are in comparison to the rest of the federal budget, to focus its assistance on those urban places most in need. For example, the tax expenditure represented by the deductibility of property taxes and mortgage interest on owner-occupied housing constitutes a housing subsidy in excess of $30 billion a year which goes disproportionately to suburban and nonmetropolitan areas. HUD's housing and urban development programs look meager by comparison and in this context, their targeting appears to begin to offset the existing anticity bias of much of the federal budget rather than to provide any unfair advantage to central cities.

The four city distress measures employed for this study are different from one another in varying degrees although they share some common components. It is not possible to judge which of the measures are more accurate or appropriate as indicators of city distress. Such a judgment must be a relative one depending on the purpose the index is to serve.

Our findings on the relationships between the city distress measures and the distribution pattern of HUD programs enable us to offer the following generalizations:

(1) Three large-scale programs, CDBG, Low-Rent Public Housing, and Section 8 are remarkably well-targeted on urban distress as measured by the Brookings Urban Condition Index and the City Needs Index; they are also targeted to some extent on distress as measured by the UDAG Impaction Ranking and the UDAG Distress Ranking.

(2) Two small-scale programs, UDAG and GNMA Targeted Tandem, are not well-targeted to the relative distress conditions among eligible cities by any of the four distress measures, although

both are highly targeted on distressed cities in the absolute. As discussed in earlier sections, neither of these programs can be adequately assessed in terms of targeting on distress with the methodology employed in this chapter. Per capita funding is an inappropriate measure for a project-specific program such as UDAG or for a demand-oriented program such as Targeted Tandem.

Thus, the current practice of HUD program distribution shows that broad-based and large-scale programs are generally well-targeted on those cities in a greater degree of distress. Ironically, some of the newer but smaller programs such as UDAG and GNMA Targeted Tandem which were created to focus HUD resources on specific targets are not closely related to the degree of distress conditions among eligible cities, despite their high degree of targeting on distressed cities in general.

Finally, while the relatively successful targeting of HUD programs might be expected to make a positive contribution to efforts to revitalize the economic, social, and fiscal conditions of distressed cities, merely knowing how well HUD programs are targeted on city distress conditions is obviously an insufficient basis for judging the extent to which those programs actually aid in retarding or reversing the decline of distressed cities.

NOTES

1. The regions used in this chapter follow the definitions of the U.S. Bureau of the Census, namely, the four main regions are the Northeast (Maine, New Hampshire, Vermont, Massachusetts, Rhode Island, Connecticut, New York, New Jersey, Pennsylvania, Delaware, and Maryland); North Central (Michigan, Ohio, Indiana, Illinois, Wisconsin, Minnesota, Iowa, Missouri, North Dakota, South Dakota, Nebraska, and Kansas); South (Virginia, West Virginia, Kentucky, Tennessee, North Carolina, South Carolina, Georgia, Florida, Alabama, Mississippi, Louisiana, Arkansas, Oklahoma, Texas); and West (New Mexico, Arizona, Montana, Idaho, Wyoming, Colorado, Utah, Washington, Oregon, Nevada, California, Hawaii, and Alaska).

2. Among the measures not considered here are: (a) the Brookings Hardship Index (unemployment rate, population under 16 and over 64 years of age, education, crowded housing, and poverty); (b) the National Planning Association Index (assessed property value, population migration, budget deficits, short-term debt, and Moody's bond ratings); and (c) Treasury Index (population change, income, city own-source revenue, long-term debt outstanding, and full market value of property).

3. A heavy concentration of leased housing in the West partly reflects the impact of a California law which requires referenda on conventional public housing. The leased housing program, therefore, is an easier alternative in California, which, in turn, affects the West's share of leased housing.

4. Only the 125 largest housing authorities were used for this study. Since the jurisdiction of public housing authorities does not always coincide with municipal

boundaries, there is often a discrepancy between actual per capita operating subsidies and the number of public housing units as a percentage of the population and the estimates presented in Table 8.6.

REFERENCES

BUNCE, H. and N. GLICKMAN (forthcoming) "The spatial dimensions of the community development block grant program: targeting and urban impact," in N. Glickman (ed) The Urban Impacts of Federal Policies. Baltimore: Johns Hopkins Univ. Press.

CHO, Y. H. (1978) "The role of the states in intergovernmental fiscal relations." Presented at the 1978 annual meeting of the American Political Science Association, New York, September.

JACOBS, S. (n.d.) "The urban impacts of the urban development action grant program." (Mimeo)

NATHAN, R. P. and P. R. DOMMEL (1977) "The cities," pp. 283-316 in J. A. Pechman (ed.) The 1978 Budget: Setting National Priorities. Washington, DC: Brookings Institution.

––– et al. (1977) Block Grants for Community Development. Washington, DC: Department of Housing and Urban Development.

President's Urban and Regional Policy Group (1978) A New Partnership to Conserve America's Communities: A National Urban Policy. Washington, DC (March).

SACKS, S. (1978) "Trends in large city characteristics: the role of annexation." (Mimeo)

U.S. Department of Housing and Urban Development (1978) The President's National Urban Policy Report. Washington, DC (August).

9

Who Benefits and Who Decides? The Uses of Community Development Block Grants

RAYMOND A. ROSENFELD

☐ PASSAGE OF THE COMMUNITY DEVELOPMENT Block Grant Program (CDBG) in 1974 brought about a major overhaul of federal efforts at urban revitalization. One of the most significant substantive issues of CDBG is the question, "Who benefits?" The issue is a controversial one because of its distributive or redistributive implications. More specifically, the concern is over the need for a community-wide urban revitalization program as against one that focuses specifically and directly upon the low- and moderate-income segments of the urban population. A second major issue of the block grant program is the procedural question, "Who decides?" A major objective of the block grant overhaul of urban revitalization programs was the decentralization of policy-making. What we would like to know is who actually decides each locality's approach to urban revitalization.

In this chapter, we shall relate these questions to urban revitalization by directing attention to a number of substantive and procedural issues addressed in the legislation, in implementing rules and regulations, and in local policy decisions.

AUTHOR'S NOTE: *Earlier versions of this chapter were presented at the annual meetings of the American Political Science Association, New York, September 1978 and the American Society for Public Administration, Baltimore, Maryland, April 1979. The author is grateful to his wife, Janelle L. McCammon, and to Donald G. Dodge of the U.S. Department of Housing and Urban Development for their assistance in preparing this chapter.*

CDBG AND URBAN REVITALIZATION

In creating Community Development Block Grants, Congress consolidated six categorical grant-in-aid programs (urban renewal, Model Cities, water and sewer facilities, open space, neighborhood facilities, and public facilities loans), thereby replacing multiple applications and program regulations with one yearly application and significantly reducing the number of pages of detailed regulations. Interlocality competition for funds for each categorical activity was replaced with a single predetermined formula-based grant. All localities of 50,000 population, central cities, and counties of 200,000 population have a legal entitlement to a designated amount of funds each year.[1] The Department of Housing and Urban Development (HUD), the implementing agency, has seventy-five days to review and disapprove an entitlement community's yearly application or the grant is automatically approved. More detailed analysis was shifted to HUD monitoring and post-auditing reviews. The overall thrust of these and other program characteristics was to shift decision-making responsibility from the federal implementing agency to local, general purpose, government officials.

Substantively, CDBG does not offer a new definition of urban revitalization. The six categorical programs that were replaced were simply folded into the block grant program. Still, the *primary objective* of the CDBG legislation sets a broad framework for urban revitalization. It states,

> The primary objective of this title is the development of viable urban communities, by providing decent housing and a suitable living environment and expanding economic opportunities principally for persons of low and moderate income [U.S. Congress, 1974: Sec 101 (c)].

This is followed by seven specific objectives that address problems of slums and blight, detrimental conditions, housing stock, community services, land use, isolation of income groups, and historic preservation. Like the primary objective, four of these specific objectives are directed to low- and moderate-income persons.

Historically there are three basic orientations to urban revitalization: (1) central city commercial and industrial development, (2) attracting middle-income residents back to the central city to live, and (3) improving low-income residential environments both physically and socially. The first two are consistent with the most widespread understanding of urban renewal, while the third is frequently identified with the Model Cities effort. In CDBG, urban revitalization is not one approach or another, but rather all three. The substantive choices available to federal officials for urban revitalization are sidestepped in favor of new procedures: let local officials select their own direction.

The three themes of urban revitalization differ substantially in the question of who benefits and when. Only the third theme, which focuses directly upon low-income neighborhoods, offers a guarantee that low-income persons will benefit. This conclusion seems to pose problems for the achievement of the low- and moderate-income focus of the primary objective of CDBG. First, the block grant program includes all three urban revitalization themes, including those that do not directly benefit low- and moderate-income persons. Second, the decentralization theme of New Federalism, in general, and CDBG, in particular, suggests that the specific approach to urban revitalization in each community should be left to local officials. Since the inception of the CDBG program in 1974, officials at the Department of Housing and Urban Development have struggled with these dilemmas: Can the CDBG program offer all three approaches to urban revitalization and operate in a decentralized manner consistent with its New Federalism roots, and still principally benefit low- and moderate-income persons?

Our approach to analyzing these issues is consistent with the notion of program implementation as "a continuation into another arena of the political process" (Rabinovitz et al., 1976: 400). The losers in congressional policy debates as well as those who fail to participate in the legislative level have another chance in the federal bureaucracy to influence the political process successfully. If they are not satisfied with the federal rules and regulations, there is always the local political arena. In fact, the policies developed at each stage influence the subsequent policies, and implementation may be viewed as an assembly process (Bardach, 1977: 37).

NEW FEDERALISM AND INTERGOVERNMENTAL RELATIONS

New Federalism was espoused by the Nixon/Ford Administrations as presenting a radical change in the relationship between the federal and state/local governments. When President Ford signed the Community Development Block Grant program into law, he heralded it for returning "power from the banks of the Potomac to the people in their own communities" (Weekly Compilation, 1974: 1060). This program was part of President Nixon's design to dismantle federal domestic program bureaucracies and federal control in favor of local policy planning and implementation (Haar, 1975: 265-268). New Federalism fits Sundquist's model of pre-1960s federal-state relations when federal funds were provided to states to achieve state goals. Sundquist (Sundquist and Davis, 1969: 4) contrasts this with an approach developed in the 1960s:

> In the newer model the federal grant is conceived as a means of enabling the federal government to achieve *its* objectives—national policies defined, although often in very general terms by the Congress.

The Community Development Block Grant program is a compromise between the New Federalism emphasis on local policy-making along with a limited role for federal bureaucracies and the replaced categorical grant approach of federal dominance. While Congress shared President Nixon's concerns with the procedural problems that arose from the categorical hodge-podge for community development, many members, particularly in the Senate, wanted a new program that would continue to emphasize substantive national objectives. The result was legislation that included substantive national objectives and some powers for the federal bureaucracy to insure that these objectives could be achieved, but local officials also were given considerable policy and implementation discretion within the national objectives.

We have already pointed out that the primary objective of the CDBG program identifies low- and moderate-income persons as the intended principal beneficiaries. This language was not included in any of the Nixon proposals or the House version but was added in Conference, based upon the Senate's community development bill. This language is juxtaposed with procedural "purposes" that emphasize the New Federalism objectives of grant simplification and local discretion. This is indicative of the overall compromise worked out in Conference that added something for everyone.

The most important language with respect to "Who benefits?" has been the requirement for certifications. The Senate included a requirement in its bill that not more than 20% of each community's budget be programmed for activities that do not directly and significantly benefit low- and moderate-income families or blighted areas. The Conference Committee dropped this numerical floor and substituted a requirement that the local government

> certify to the satisfaction of the Secretary that its Community Development program has been developed so as to give maximum feasible priority to activities which will benefit low- or moderate-income families or aid in the prevention or elimination of slums or blight . . . [or that meet] other community development needs having a particular urgency [U.S. Congress, 1974: Sec. 104 (B) (2)].

In bureaucratese, this is called the "max feas test" or the "three prong test."

The idea of a certification is consistent with the procedural goal of increasing local policy planning and implementation responsibilities and decreasing the role of the federal bureaucracy. A certification, which is a promise of regulatory compliance as opposed to a priori proof, gives wider discretion to local officials. Certainly, it lessens the involvement and control of the federal bureaucracy at the application stage of a grant program. The substance of the certification is a restatement of congressionally determined national objectives. Thus, the *notion* of a certification is consistent with the local flexibility of the older model of federalism, while the *substance* of the "maximum feasible priority" certification reflects an effort to achieve national goals similar to the newer model developed in the 1960s.

Of particular interest in the certification language is the use of the "maximum feasible" phrase. These words are reminiscent of the "maximum feasible participation" concept of the Community Action Program (CAP) of the 1960s (Moynihan, 1969). At that time, the concept did not have a precise legislative history, and, as a result, it became the basis for considerable misunderstanding. The legislative history of the block grant program is also silent about the meaning of "maximum feasible priority." It appears to be no more than convenient rhetoric designed to achieve a compromise between urban revitalization goals that emphasize benefit for low- and moderate-income persons and other substantive objectives of CDBG and the decentralization orientation of New Federalism.

Inherent in a legislative compromise that includes national objectives, on the one hand, and only a vague federal implementation tool such as the "max feas" certification, on the other, is the creation of wide administrative discretion to seek out the "true meaning" of the legislative language. The primary objective of the block grant program states that low-*and* moderate-income persons shall be the principal beneficiaries of these funds, while the certification makes benefits for low- *or* moderate-income persons only one of three program components (along with the prevention or elimination of slums or blight, and meeting urgent community development needs). The history of the legislation does not explain the intention of the "low *and* moderate" versus "low *or* moderate" terminology, nor the relationship between the apparently more limiting primary objective and the broader certification language. Finally, the legislation does not include a definition of "low- and moderate-income." As a result, Frieden and Kaplan (1975: 251) hypothesized that the compromise on a "maximum feasible priority" certification to achieve the national objective of principally benefiting low- and moderate-income persons would lead to "maximum feasible misunderstanding" as it did in the decade earlier.

BUREAUCRATIC FLEXIBILITY

Federal bureaucrats play an increasingly important role in program implementation. Rabinovitz et al. (1976: 400) point out that the job of a federal executive is "to turn legislation into workable practice by balancing the claims of legislative intent, public opinion, and administrative effectiveness." These bureaucratic considerations together with CDBG legislative characteristics such as a national substantive objective to benefit principally low- and moderate-income persons, a vague federal instrument (the "max feas" test) to assure that this objective would be achieved, and procedural objectives of increased local policy-making and implementation responsibilities ultimately led to increased bureaucratic power to make broad interpretations of the CDBG legislation. Further, it could be expected that the steady direction of these interpretations would be toward the national objectives and away from the procedural issue of decentralization. Increasingly, urban revitalization would be limited to the third approach previously cited—activities that improve low- and moderate-income neighborhoods and provide direct benefits to their residents.

HUD has been writing rules and regulations and administrative notices for the CDBG program since the fall of 1974. There appears to be a clear movement from fewer words to more words, from local discretion within broad program parameters to local discretion within specific standards for the pursuit of national objectives, and from minimal application review to substantial front-end analysis. The movement has taken place during a Republican and a Democratic administration, but has quickened during the Democratic reign. These changes followed from the way the New Federalism objective—to decrease federal intrusion into local decision-making—was compromised during the legislative process. HUD critics, during the initial years of CDBG, complained that HUD was ignoring the national objectives of the legislation. This is balanced by the HUD critics during the current Democratic Administration who complain (with equal force) that HUD is overstepping its legitimate authority in establishing precise program standards to implement the "who benefits?" language. Until Congress changes the law or gives itself a veto over HUD policies (which has recently occurred), the power to establish a national policy toward urban revitalization and the "true meaning" of New Federalism with respect to CDBG rests with HUD officials.

The original program regulations (U.S. Federal Register, 1975) promulgated in late 1974 and early 1975 showed the determination of HUD officials to minimize federal intrusions into local decisions. The regulations were about as close to a simple reprint of the legislation itself as possible. The primary objective of the act and the local certification for "maximum feasible priority" were repeated without embellishment, implementation guidelines, or performance standards. In no way were grant recipients required to act on the primary objective of the entire CDBG program in setting local policies. The certification itself was one of many included on a government-printed application; when the local chief executive signed an application, he fulfilled the requirement for "maximum feasible priority" and was free to pursue any approach to urban revitalization the community selected. The regulations did not emphasize one "prong" of the certification over the others, and no guidelines were written upon which the certification could be questioned. At this point, the answer to the "who benefits?" question was based almost entirely upon local decisions.

Some will maintain that in accordance with the legislative compromise on national objectives and the decentralization theme of New Federalism, HUD had merely shifted the federal review of national objectives to the performance stage. The regulations concerning "secretarial review and monitoring of recipient's performance," however, contain no reference to the "who benefits?" issue. Further, the Comptroller General (U.S. General Accounting Office, 1976: 10-11) found HUD field staff held widely divergent views of the meaning of "maximum feasible priority," ranging from "no meaning at all" to "51 percent of the funds must benefit lower-income census tracts or aid in preventing or eliminating slums or blight." Because no field staff actions on this language were required in the initial years, the misunderstanding had little *consistent* impact on the CDBG program. The legislative language gave HUD tremendous flexibility. Consistent with the New Federalism philosophy of decentralization, this flexibility was used by HUD's Central Office to grant local officials maximum discretion. Some field offices, however, began to use the flexibility to achieve a particular approach to urban revitalization.

The creep toward centralized power and a national approach to urban revitalization seemed inevitable. Although in the beginning (at least in terms of regulations) there was a major effort to emphasize New Federalism themes, the legislation created too many vague and inconsistent program characteristics that would ultimately need to be reconciled. Administrative regulations could only become more detailed, complex, and rigid in defining the parameters within which local officials could work.

HUD Assistant Secretary Meeker was aware of the pressure for more federal involvement with respect to implementation of the CDBG national objectives. In an April 26, 1976 statement during oversight hearings before the Senate Committee on Banking, Housing, and Urban Affairs, Mr. Meeker (1976a) pointed out: "This statutory certification requirement is less simple and straightforward than may first appear, and has often been misunderstood. After all, what does *maximum feasible priority* mean, and in whose eyes? . . . I expect that maximum feasible priority will continue as an issue for some time in the future." It is not that he was unable to define "maximum feasible priority" but rather that to do so would entail a national effort to set specific guidelines for urban revitalization.

By May 1976, Meeker was forced to send a clarification memo to his field staff on the issue of "maximum feasible priority." He pointed out that all activities must meet one of three components of the certification language (benefit low- or moderate-income persons; aid in the prevention or elimination of slums or blight; or meet an urgent community development need). Yet, still consistent with New Federalism themes, he added: "The Department is not to make a judgement between activities that would meet the standards. The Department is to accept the certification to the program as a whole" (Meeker, 1976b). Thus, each activity included in a local community development program must fulfill the certification requirements, but HUD did not care which one of the three. It was necessarily vague, since HUD had not defined what it meant to benefit low- or moderate-income persons or to eliminate slums and blight.

However small this step was in the minds of those in HUD's Central Office, it was a major change for many of the field staff. In effect, it gave the field staff an opportunity to review each individual activity against the certification language. In order to do this, some standards were needed, and some field personnel began to delve into the substance of activities and to ask formal questions of local community development officials. If misunderstanding followed, it was not because of the legislative language but rather because HUD chose to stay away from precise national program standards.

The slow movement toward detail and increasing federal requirements continued into the last days of Republican leadership at HUD and for two months after Jimmy Carter was elected President. Twenty-six months after the initial program regulations for CDBG were published, HUD required that each activity in a local Community Development Plan must pass the three-prong test (U.S. Federal Register, 1977a: 6504-5). The "max feas" language was transformed from a benign certification to an active program requirement. Rather than have HUD staff look behind the certifications, as Meeker had suggested, the burden shifted to the localities to indicate the relationship of each specific project included in an application for funds to the certification language. The major changes were an increase in local requirements, a move from broad program certifications to specific activity designations, and a shift of burden from the "feds" to the localities.

Long before the change in administrations at HUD, pressures were building for the development of specific implementation criteria by which to judge performance in terms of "who benefits?" Thus, the Comptroller General's Report of June 23, 1976, requested by the Chairman of the Senate Subcommittee on Housing and Urban Affairs, recommended that the Secretary of HUD

> define and develop criteria for determining maximum feasible priority to be used by communities in preparing their Community Development Programs and by area offices in evaluating and monitoring the programs and . . . determine, during application review, whether an applicant's program meets the maximum feasible priority criteria [U.S. General Accounting Office, 1976: 20].

Evaluation and monitoring studies were published at about this time by the Southern Regional Council (1976), the National Association of Housing and Redevelopment Officials (1976), and HUD's Office of Evaluation in Community Planning and Development (1975). Each of these studies reported problems in the implementation of the block grant program, ranging from low percentages of funds being programmed for low- and moderate-income neighborhoods to flagrant violations of the "intention" of the CDBG program such as funding bike paths and tennis courts in middle-income neighborhoods.[2]

Within three months of the new Democratic administration, HUD sent an administrative notice to its field staff and all CDBG recipients outlining a new "maximum feasible priority" policy (Embry, 1977). The authorizing legislation had not yet changed, but the new top officials at HUD saw a need for more specific criteria to guide local policy and implementation decisions. The new policy on April 15, 1977 reaffirmed the last Republican regulations requiring localities to develop a Community Development Program in which each proposed activity meets the so-called "three-prong" test of the certification. More importantly, for the first time, the notice set national definitions for each of the three components of the certification. Activities were described in some detail. Now HUD and the localities would have a basis upon which to determine the validity of local certifications. The notice stated, "If a proposed activity cannot be shown to meet one of these objectives, it should not be approved" (Embry, 1977: 3). This was another step toward a national approach to urban revitalization.

The new policy continued an important legislative interpretation made first by the Republicans at HUD. This interpretation basically stated that the program certification for "maximum feasible priority" in Section 104 (B) (2) of the Act—the three-pronged test—was the only legislative reference to "who benefits?" that might be used as a program requirement for local officials. Thus, Section 101 (C) which states the primary national objective and directs the entire block grant toward activities "principally for persons of low- and moderate-income" was viewed as rhetoric only. (New HUD officials continued to shy away from defining "principally" for program purposes.) The new notice would "guarantee" that all CDBG funds were to be spent consistent with the program certification, *but no one component of that certification would be given preference over the others,* thus maintaining wide local discretion in selecting urban revitalization strategies.

It is interesting to note that the April 15, 1977 policy notice was sent out only two months after the new Democratic Assistant Secretary for Community Planning and Development (CPD) came to HUD. The notice was the result of limited consultation outside HUD and a heavy reliance upon HUD staff who wrote CDBG program regulations under the Republicans. From the passage of the CDBG legislation in August 1974, top HUD officials had been saying that the primary national objective of the act was not and should not be a program requirement. Therefore, even if HUD were to define the components of the program certification, the primary national objective could not be used to give greater weight to activities that benefit low- and moderate-income persons over other activities. Furthermore, the New Federalism origin of CDBG dictated limited federal intrusion and maximum local flexibility. While the April 15, 1977 notice was important in strengthening HUD's hands and eliminating abuse, it did not mark a major departure from Republican policies.

Upon issuance of the April 15, 1977 notice, HUD embarked upon the development of comprehensive new program regulations for CDBG. A central issue was targeting funds *within* block grant communities to assure that low- and moderate-income persons benefited from the program, thus limiting the available approaches to urban revitalization. HUD's attempt at rewriting regulations is an excellent example of program implementation as the "continuation into another arena of the political process" (Rabinovitz et al., 1976: 400).

The participants in the development of new program regulations were many. From outside HUD the central actors were members of a Working Group for Community Development Reform—a coalition of neighborhood, civil rights, open housing, labor, and other organizations. The Working Group prepared a detailed analysis of the CDBG legislation and program regulations for the new Democratic Administration and made very specific recommendations for changes. Their goal with respect to "who benefits?" was to target more funds to low- and moderate-income persons, consistent with the primary national objective of the act. Their proposal was to overlay that language onto the program certification, thus, providing a legislative basis for administrative policies establishing minimal percentages of funds to be spent for low- and moderate-income persons. This group was the contingent in favor of emphasizing national objectives and establishing precise tools for HUD review of local policy choices. Ultimately, the coalition had a "designated" spokesperson within HUD—a special assistant to the assistant secretary for CPD, selected from this group to reflect their point of view.

The opponents to a shift away from the New Federalism emphasis of CDBG were disorganized and ill-prepared to defend the status quo. They consisted of the "city folks"—the U.S. Conference of Mayors, the National League of Cities, and the National Association of Counties, among others. Representatives from large and small cities were united in their opposition to an increased federal role in local decision-making. Some of these officials were in an awkward position given the issue and their constituencies. While a big city mayor might naturally prefer an emphasis on local flexibility rather than federal performance standards, the issue would have consequences for low- and moderate-income constituents of many Democratic mayors. To be outspoken in opposition to changes could have been interpreted as working against one's constituents. This group, too, was "represented" in HUD with a special assistant to the assistant secretary.

The third major cluster of participants was the HUD bureaucracy. It is inaccurate, however, to refer to this as a "group," for there were about as many opinions on the "max feas" issue as there were HUD personnel.

The battle lines over the regulations were drawn as soon as the April 15, 1977 notice was published. Meanwhile, Congress was considering reauthorization of the CDBG program along with a number of amendments dealing with the same issues being addressed at HUD. HUD officials were deeply involved in the legislative changes at the same time they were preparing new program regulations. In October 1977, Congress passed and President Carter signed into law the Housing and Community Development Act of 1977 containing some important amendments to the original CDBG legislation with respect to "who benefits?" (U.S. Congress, 1977a). These legislative changes came to the 1977 Act through Senate Bill S. 1523. The intent of the new language is clearly stated in the Report of the Committee on Banking, Housing, and Urban Affairs of the Senate (U.S. Congress, 1977: 11), that accompanied S. 1523: "The Committee bill would include in Section 103 a series of amendments to relate, more closely, the program requirements of the statute to the statutory objectives of benefiting low- and moderate-income persons." The Senate changes that were included in the final Act amended the "max feas" certification to read "low- *and* moderate-income" rather than "low- *or* moderate-income" in order to clarify the intent that both groups benefit from CDBG expenditures. Although the primary objective stated that CDBG funds should principally benefit low- and moderate-income persons, the certification continued to be the operable language. One implementation study (National Association of Housing and Redevelopment Officials, 1977) of the block grant program indicated that few low-income neighborhoods were receiving funds that benefited the residents. The "low- or moderate-income" certification language had provided the legislative basis for such local programming decisions.

The 1977 Amendments also clarified the basis upon which HUD may disapprove a locality's application for entitlement funds to include specifically problems with the "max feas" certification and its three-prong test. Even with these changes, the Senate did not propose and Congress did not include specific performance standards, nor a direct basis for giving greater emphasis to one of the three prongs of the "max feas" certification (namely benefit for low- and moderate-income persons) over the others.

Other substantial changes in the CDBG program included in the 1977 Amendments were:

- a new needs formula for entitlement grantees to direct more funds to distressed communities,
- a new focus on economic development projects, and
- more specific citizen participation requirements.

After many internal drafts of new regulatory policies, HUD published proposed regulations in October 1977 that for the first time involved implementing the primary national objective of the CDBG legislation (U.S. Federal Register, 1977b: 56466). They proposed that at least 75% of each community's CDBG funds (excluding planning and administrative costs) be budgeted for activities that principally benefited low- and moderate-income persons. A system for calculating funds that contribute to the 75% figure was also presented. The impact of these decisions was to place one particular approach to urban revitalization above all others.

While the outcry from local officials, Congress, and public interest groups was considerable, the new standards were published intact in final form on March 1, 1978 (U.S. Federal Register, 1978b). Some changes were made in calculation methods and requirements for documentation, but HUD kept the 75% standard. The concept of a certification was maintained. HUD defined "maximum feasible priority" as well as the three prongs of the test, and created a basis for double-checking local decisions and for conducting performance reviews after the fact.

These new guidelines say that "maximum feasible priority" is achieved if each funded project passes the newly defined "three prong" test and if at least three-quarters of a community's CDBG funds are programmed to benefit low- and moderate-income persons.[3] The remaining one-quarter of the grant could be used to eliminate slums and blight or meet other urgent needs.

Projects that benefit low- and moderate-income persons include those with eligibility requirements limited to low- and moderate-income persons, those benefiting a designated neighborhood where a majority of the residents are of low and moderate income, and those which create jobs primarily for low- and moderate-income persons. Some exceptions to these policies were included where low- and moderate-income persons are not concentrated in a few neighborhoods.

The second prong of the test includes projects which prevent or eliminate slums or blight. This includes activities "in an area which is a slum, or a blighted, deteriorated or deteriorating area as defined by State or local law and in which a comprehensive program ... to remedy the conditions which qualify the area as an urban renewal or similar area [is planned]." Also included are "projects designed to eliminate detrimental conditions which are scattered or located outside slum or blighted areas" and completion of federally assisted urban renewal projects.

The third prong of the test is for urgent community development needs, narrowly defined as projects "designed to alleviate a serious and immediate threat to the health or welfare of the community which is of recent origin where the applicant is unable to finance the projects on its own, and other sources of funding are not available."

While this new policy continues the "maximum feasible priority" certification, it relies upon the primary national objective to give greater importance to one of the three certification "prongs" than the others. Very clearly and very deliberately, HUD placed severe limitations on the available strategies for urban revitalization.

In spite of or because of the legislative changes, some maintained that HUD's actions were inconsistent with the law and the local flexibility theme of CDBG. The HUD response was included in the preface to the March 1, 1978 regulations: "Our intention is to carry out the statutory objective of benefiting low- and moderate-income persons in a strong and committed fashion. At the same time, we intend to be practical and flexible regarding documentation." The opportunity for bureaucratic centralization was present in the original legislation, and it was inevitable that it would be exercised to implement the national objectives of CDBG.

Within two months HUD began to back down somewhat. A new policy notice (Embry, 1978: 6) was sent to field staff on April 28, 1978, in which the percentage requirement for funds targeted to low- and moderate-income persons was lowered to 50%. If a community certifies that 75% or more of its grant will benefit low- and moderate-income persons, HUD will limit application review; if the application proposes 50-74%, field staff will conduct detailed reviews, but may still give approval; if less than 50% of program funds principally benefit low- and moderate-income persons, only the Central Office can give approval.

Even with this policy change to a reduced percentage requirement for low- and moderate-income benefit members of the House of Representatives were upset with HUD's movement away from the New Federalism emphasis on local flexibility (Congressional Quarterly Weekly Report, 1978). HUD had placed activities which benefit low- and moderate-income persons on a plane above the elimination of slums and blight or urgent community development needs. This direction had been rejected by Congress in the 1977 legislative amendments although the primary national objective was still in conflict with the certification language.

In October 1978 a new set of Housing and Community Development Amendments was passed by Congress which attempted to address the problem of bureaucratic regulations that placed substantial limitations upon urban revitalization strategies and local policy-making discretion. The new legislation states:

> The Secretary may not disapprove an application on the basis that such application addresses any one of the primary purposes described in [the maximum feasible priority certification] to a greater or lesser degree than any other [U.S. Congress, 1978].

There is an "out" for HUD, however:

> Such application may be disapproved if the Secretary determines that the extent to which a primary purpose is addressed is plainly inappropriate to meeting the needs and objectives which are consistent with the community's efforts to achieve the primary objective of this title [U.S. Congress, 1978].

But to make sure that HUD would not overstep its bounds, Congress (1978: Sec. 324) mandated that all future administrative rules or regulations be submitted for a potential congressional veto before such rules were implemented.

In January 1979 HUD drafted a new notice to its field staff to guide future application reviews. Still using the flexibility derived from the original legislative conflicts among urban revitalization strategies and decentralization themes, HUD held its ground on the special place for activities which benefit low- and moderate-income persons. An application will be reviewed to determine if the proposed projects are "plainly inappropriate to meeting the needs and objectives identified by the applicant" if less than 75% of funds planned for the current year and for a three-year period are budgeted for low- and moderate-income persons. Field staff may still approve an application unless the one-year or three-year plan calls for less than 50% of a community development budget to benefit low- and moderate-income persons. If this is the case, only Central Office personnel at HUD can approve an application.

Clearly HUD's position on urban revitalization has changed substantially since 1974. Originally the approach was that any urban revitalization strategy selected by local officials was acceptable. More recently HUD has directed local officials toward strategies that principally benefit low- and moderate-income persons, yet an application may be accepted by field staff if only 50% of a grant is for this purpose. Ostensibly, it appears that HUD has limited local flexibility in favor of a national urban revitalization policy. Unfortunately, it is not clear if this national policy originates in Congress or the bureaucracy at HUD.

Because there is no final answer to the claims of legislative intent, public opinion, public interest group dominance or administrative effectiveness, the inherent vagueness of the "max feas" language results in maximum feasible *flexibility* for the bureaucracy at HUD. The "correct" understanding is whatever the federal bureaucrats say it is, and this will vary. The bureaucracy may choose to rely on imprecise language to interpret "max feas," and misunderstanding may result; but the bureaucracy may also rely on very precise program standards, in which case a clear understanding may follow. Thus the "max feas" language gives the bureaucracy wide discretion.

LOCAL IMPLEMENTATION

Up to this point, this chapter has analyzed the output of the CDBG program with particular attention being given to implementation decisions for urban revitalization and who *should* benefit. From this point forward, we shall focus upon the outcomes of Community Development Block Grants: Who does benefit from these federal funds?

With the emphasis of New Federalism upon procedural issues such as local discretion in policy-making and program implementation, the key substantive outcome decisions are supposed to be made by local officials. Although the CDBG legislation and regulations have fluctuated in the degree of emphasis on certain desired outcomes, local officials have maintained their authority to direct the *actual* outcome of their CDBG programs. The legislative and regulatory issues with regard to "who benefits?" have never really strayed from the New Federalism framework of broad local discretion. Rather, these issues have been concerned with the parameters within which local discretion is to be exercised. At no time has Congress or HUD proposed, for example, that the CDBG program be directed exclusively to one particular urban revitalization strategy. During the Republican Administration when the "maximum feasible priority" certification was unembellished, as well as during the current period of more precise implementation standards, local officials have responsibility for the ultimate choices. Arising from this characteristic of New Federalism, in general, and CDBG, in particular, is the widespread criticism that decentralization works against the interests of the poor. It is this criticism that directs our analysis to the outcomes of the CDBG program.

One type of decision is made at the congressional level: the development of a statistical formula for the distribution of funds among local governments. This was a central issue in 1974 when the original CDBG legislation was written, and continued to receive considerable attention through the 1977 amendments.[4] Although formula decisions are judgmental and have to do with winners and losers (i.e., local governments), they are the basis for distributing funds among localities and not *within* localities. A city receives funds based upon an overall needs formula, but federal policies do not clearly require that the funds be spent to alleviate problems symbolized by the formula variables. Such decisions are left to local discretion. Thus, the decisions that have consequences for people and a local community are still made by local officials when they decide how to spend CDBG funds within the broad parameters set by legislation and regulations.

Most studies of CDBG outcomes have been concerned with area-benefit activities using what Brookings (Liebschutz, 1977; and Nathan et al., 1977b) calls the *sociospatial* perspective. This methodology determines where an activity occurs (what census tract) and associates benefit with service area demographics (also defined as the census tract). "Low-" and "moderate-income" is defined as a census tract whose median income is less than 50% and 51-80%, of the median income of the area. Most studies including HUD's own annual reports on CDBG define area as Standard Metropolitan Statistical Area (SMSA), although some use the city median income. Finally, this methodology assumes that 100% of a project benefits low- and moderate-income persons if the median income of the census tract in which the project is located is less than 80% of the median income of the area. Major evaluations of CDBG by HUD (1975, 1977, 1978), the Brookings Institution (Nathan et al., 1977a; Dommel et al., 1978), the

National Association of Housing and Redevelopment Officials (1976), and the General Accounting Office (1976) have all pursued the sociospatial analysis of "who benefits?" from CDBG expenditures in spite of some basic methodological shortcomings.

A major problem with the sociospatial approach is the occurrence of both under- and over-estimates of "true" benefit. This problem arises from four choices within the methodology as commonly used. First, the census tract may not accurately reflect the service area of a project. For example, a park may be located in a census tract whose residents do not have access to the facility as a result of an expressway on its border. Second, direct-benefit projects are treated like area-benefit ones. A housing rehabilitation loan or grant program may be limited to persons whose incomes are less than 80% of the median income of the area. Yet, the loans could be directed to neighborhoods that are not primarily low- and moderate-income. The sociospatial methodology would not allocate the cost of this rehabilitation project to the low- and moderate-income benefit category. Third, the choice of city or SMSA as the definition of area will expand or limit the number of census tracts that qualify as low and moderate income, depending upon the characteristics of the metropolitan area. Fourth, allocating 100% of a project's cost to a particular benefit category based upon a single cut-off point obscures the benefit data. A neighborhood facility may have a particularly dense service area or be located in a dense census tract where the median income is just above the cut-off point for being classified as low- or moderate-income benefit. In fact, there may be significantly more benefit derived for low- and moderate-income persons than in a similar project in an exclusively lower-income neighborhood with fewer residents.

It is this writer's opinion that most studies will continue to rely heavily upon the sociospatial approach to measuring benefit from CDBG-funded projects because it is simple and minimizes judgmental calls by the researcher. A precise standard is used and one assumes that the overestimates cancel out the underestimates. For the sake of simplicity of analysis, we tend to shy away from questions of horizontal equity (are all lower-income areas receiving equal treatment?). With the relatively low level of expenditures for community development in relation to the level of need, questions of horizontal equity are a luxury. Clearly, all neighborhoods in need of federal development assistance are not receiving equal treatment.

Shortly after the 1974 Housing and Community Development Act was signed into law, HUD began a major data collection effort as part of its legislatively required annual report to Congress on the CDBG program. A stratified sample of 151 metropolitan entitlement grantees was selected to make possible a thorough analysis of local policy decisions. The sample size was determined by an optimum allocation formula at the 95% confidence level with a 5% sampling error (Arkin, 1963: 196). The universe of

792 first-year entitlement communities was stratified according to entitle-ment grant amount (over $4 million, $1-4 million, and under $1 million).[5]

HUD has been collecting these data for each year of the CDBG program,[6] and this outcome analysis is the basis for their estimates of the percentage of funds that is being spent primarily to benefit low- and moderate-income persons. Table 9.1 presents a summary of the "who benefits?" findings for the first, second and third years of the CDBG program. These data rely upon the SMSA for the definition of "area."

The first-year analysis of CDBG found that local officials developed plans for spending 69% of program funds in low- and moderate-income neighborhoods. Given the absence of HUD-determined standards for the primary beneficiary national objective, this first-year figure appears to be remarkably high. During the second and third years of the program, however, there was a major decline in the percentage of funds directed to the primary beneficiaries of the act; most of the decline occurred between the first and second years. The analysis for the third program year indicated that 61.7% of funds was directed to low- and moderate-income neighborhoods.

As HUD debated the primary beneficiary issue during the Republican Administration and made some adjustment in the program regulations, it appears that local officials who originally read the legislation as an attempt to improve lower-income neighborhoods or who were busy completing urban renewal or Model Cities plans during the first year of CDBG began to exercise their discretion to work throughout their communities. This, of

TABLE 9.1   Who Benefits from CDBG Funds?
(percentages)

Income Group	Fiscal Year		
	1975	1976	1977
Low and moderate[a]	69.0	62.1	61.7
Low	b	(11.3)	(10.5)
Moderate	b	(50.8)	(51.2)
Other	31.0	37.9	38.3
Total	100.0	100.0	100.0

a. Low and moderate income are defined as census tracts whose median income is less than 50 and 51-80%, respectively, of the median income of the SMSA.
b. Not available.
Source: First, Second, and Third Annual Reports on the Community Development Block Grant Program, Housing and Community Development Act of 1974, Department of Housing and Urban Development, December 1975, June 1977, and March 1978.

course, is consistent with the decentralization theme of CDBG, but potentially inconsistent with the "who benefits?" national objective. Between the first and second program years, local governments expanded the number of census tracts receiving CDBG funds from 35.8% of all tracts to 53.0% (Rosenfeld, forthcoming). With this expansion came a movement into lower-income areas that had been excluded under the replaced categorical programs as well as many higher-income neighborhoods. Local discretion was used to pursue a variety of urban revitalization strategies and to satisfy a variety of constituencies in each local community.

We have indicated that there are substantial methodological problems with the sociospatial approach pursued with these data. While we do not yet have the ability to overcome these problems at the aggregate level, we may look at the types of activities being planned for low- and moderate-income neighborhoods and develop a better understanding of the benefit data thus far presented.

Table 9.2 presents an analysis of the types of activities being funded in the low- and moderate-income neighborhoods for the first, second and third years of the Community Development Block Grant program. Substantial changes have occurred in local plans during this period. In particular, clearance-related activities, such as land acquisition, relocation, and demolition, that began as almost one-half of CDBG expenditures in the lower-income areas declined rapidly to 23.8% by the third year. Expenditures for housing rehabilitation also increased from 13.6 to 19.9% of the expenditures in the low- and moderate-income neighborhoods.

TABLE 9.2   Funded Activities in Low and Moderate Income Neighborhoods (percentages)

		Fiscal Year	
Activity	1975	1976	1977
Clearance-related	43.4	32.1	23.8
Code enforcement	2.2	1.3	1.5
Other public works	14.1	19.7	28.7
Water snd sewer	4.3	1.4	1.3
Open space and neighborhood facilities	5.2	7.8	7.3
Housing rehabilitation	13.6	21.2	19.9
Service facilities	6.5	3.9	4.9
Public services	10.7	12.5	12.5
Total	100.0	100.0	100.0

Source: First, Second, and Third Annual Reports on the Community Development Block Grant Program, Housing and Community Development Act of 1974, Department of Housing and Urban Development, December 1975, June 1977, and March 1978.

The question of the primary beneficiaries must be raised in conjunction with a number of these planned activities. First, the land acquisition, relocation, and demolition activities, grouped together as "clearance-related" are the traditional mainstay of the urban renewal program. Such projects usually include substantial expenditures for public works after the land has been acquired, and this is the pattern seen in these data. After an initial burst of clearance-related activities, local policy shifted to public works activities. The typical image of urban renewal is the achievement of massive land use changes from lower-income residential to higher-income residential, commercial, or industrial. As noted earlier, most direct benefits from urban renewal do not go to low- and moderate-income groups. Although, as Sanders's chapter illustrates, this changed over time. Thus, some of these urban renewal activities were undoubtedly from Neighborhood Development Programs of the late 1960s and early 1970s, in which case residential rehabilitation, spot demolition, and improvement of the infrastructure through public works activities may have been more of a programmatic thrust. In such cases benefits may have accrued directly to low- and moderate-income residents. Given the history of clearance-related activities, however, it seems reasonable to conclude that while these projects are in lower-income neighborhoods, they do not *all* benefit primarily the low- and moderate-income residents.

The question of real accrued benefit for lower-income residents may also be raised for the substantial amount budgeted for housing rehabilitation loans and grants. A widely touted (but numerically small) demographic movement in the United States is the return of the middle class to the central city. HUD's own Urban Homesteading program is designed in part to support this movement. Cities like Baltimore, Minneapolis, Washington, DC, and St. Louis are devoting considerable public funds to support the return of the middle class to the central city. A growing concern, however, is the potential and real displacement of lower-income individuals from their central city neighborhoods. Thus far, HUD has not limited direct-assistance CDBG programs to low- and moderate-income individuals. While most cities are conducting their rehabilitation programs in lower-income areas, it is not at all clear that the loans and grants are directed to the current residents. One problem that might be expected is that lower-income homeowners may be hesitant to undertake additional debt to rehabilitate their residences, while many local officials are hesitant to fund rehabilitation grant programs exclusively. There is a better than reasonable chance that a sizable portion of the funds for housing rehabilitation in lower-income neighborhoods is not benefiting the lower-income target group mentioned in the original CDBG legislation.

Finally, the activities data for low- and moderate-income neighborhoods include substantial funds for public services. Because most public services, by nature, distribute direct benefits to individuals (i.e., day care, elderly homemaker assistance, health care, etc.), it is possible to determine precisely who benefits. The reality, however, is that most cities report on the characteristics of the service area and limit their public services primarily to the lower-income neighborhoods. The extent to which higher-income residents benefit from these activities is unknown. The nature of the activities, however, may suggest that the lower-income residents of a service area receive their "fair share" of the benefit. Many of these services are similar to those funded through state, local, and federal welfare efforts.

This brief discussion of the planned activities in the low- and moderate-income census tracts during the first three years of the CDBG program supports the conclusion that the total percentage of funds directed to benefit principally low- and moderate-income persons could be considerably lower than the figures cited earlier in this chapter.

CONCLUSION

One of the apparent goals of New Federalism was to substitute a new decentralized decision-making procedure in place of specific national substantive policies in such areas as housing, community development, education, law enforcement, and social services. Somewhere along the road to legislative enactment, however, the Community Development Block Grant Program compromised this New Federalism goal. Substantive national objectives were included in the legislation along with a vague and potentially conflicting implementation device for their achievement. These compromises set the stage for widespread federal bureaucratic discretion to determine the "final" meaning of the legislation. The impact of this is not only to create a situation in which federal bureaucrats may set the parameters within which local policy making occurs, but to guarantee the ever-narrowing of these parameters. This is a far cry from the original decentralization theme of New Federalism.

The CDBG program offers local officials many approaches to urban revitalization, but the inclusion of a national objective principally to benefit low- and moderate-income persons inherently limits the available strategies. Projects designed to revitalize downtown business districts and to attract middle-income persons and commercial and industrial facilities to the central city may be inconsistent with the "who benefits?" national objective. It is not at all clear that Congress intended to place this limitation upon the program.

Political compromises in Congress must have some consistent theme to them, or bureaucratic red tape is destined to increase. Including conflicting passages from a House and a Senate bill and ignoring the consequences are not constructive. Decentralizing decision-making and decreasing the role of the federal bureaucracy cannot be achieved if Congress creates unnecessarily vague and conflicting objectives and implementation devices. Ultimately, the bureaucracy will be forced to reconcile the conflicts thus expanding their decision-making role.

During the Ford Administration, the Republicans sought to emphasize the procedural objective of local discretion over the national substantive objective for the primary beneficiaries of the CDBG program. The local response was to direct substantial portions of CDBG funds into low- and moderate-income neighborhoods. Whether or not these funds benefited the current population of these areas is in question given some basic problems with the sociospatial methodology for measuring benefit.

The change from a Ford to a Carter Administration presented an opportunity for federal officials to exercise their freedom to define more carefully the parameters within which local officials must work. Very quickly, the Democrats moved to emphasize the national substantive objectives of the block grant program and to establish performance standards to assure that the primary beneficiaries would be low- and moderate-income individuals. There is some support for the conclusion that a new set of problems was created. The most recent program standard for the percentage of funds that must be spent to benefit low- and moderate-income individuals is actually less than the level of funds budgeted for this group by local officials prior to implementation of the new standards. A further decline in funds budgeted for low- and moderate-income persons could result.

Because the newest policies rely heavily, but not exclusively, upon sociospatial concepts, we may find that in the name of clear and precise program standards, the Carter Administration has given local officials greater flexibility than they were willing to exercise on their own. It is not that these regulations give greater flexibility than the Republican approach but that the shift from vagueness to specificity in national policies makes it possible for local officials legitimately to *exercise* more flexibility than they had in the past. In effect, the newest regulations may have opened up urban revitalization strategies that local officials thought were beyond the intent of CDBG.

During the 1960s a number of mayors were said to be attracted to the Model Cities program in contrast to a variety of alternatives that would have given them greater flexibility (Frieden and Kaplan, 1975: 219). Local officials wanted to see more limitations placed upon their discretion:

And as Mayor Landrieu of New Orleans could point out, model cities allowed him to play "catch-up" in the deprived areas of his city. It gave him a political defense; for he could tell the community, "Look, this money is here to be spent and spent only in these particular neighborhoods [Haar, 1975: 265].

While the Democrats' new CDBG policies recognize the need for a political defense at the local level, they have established one that in terms of benefiting low- and moderate-income persons may be less effective than no defense at all. The ambiguity during the Republican days gave some local officials a defense: they could argue whichever side they wanted and were assured of being "right" on the principal beneficiaries of their CDBG program. Being right is now clear: local officials may spend one-half of their CDBG funds for activities that do not principally benefit low- and moderate-income individuals. It may soon appear that local discretion in the midst of national ambiguity is preferable to national performance standards that are set too low. Maximum feasible misunderstanding may be functional for the "who benefits?" question, particularly in comparison to a clear understanding that offers excessive flexibility.

NOTES

1. A more detailed analysis of the major features of the CDBG program may be found in U.S. Congress (1975).
2. As might be expected, HUD's own evaluation reports were descriptive without drawing evaluative conclusions.
3. In February 1978 HUD published regulations which defined low- and moderate-income as "families whose incomes do not exceed 80 percent of the median family income of the metropolitan area, or in the case of families residing in non-metropolitan areas, of the non-metropolitan areas of the state" (U.S. Federal Register, 1978a). This was the first time HUD defined what "area" actually meant.
4. The original formula contained in the 1974 legislation distributed CDBG funds based upon poverty population, overcrowded housing, and population on a 2:1:1 ratio. The 1977 amendments added an alternate formula to direct additional funds to distressed communities. It consists of growth lag, poverty population, and age of housing on a 2:3:5 ratio. For a full discussion of these formulae, see Nathan et al., 1977a; Dommel et al., 1978; and U.S. Department of Housing and Urban Development, 1978. Also see the Ross chapter in this volume.
5. A more detailed explanation of the sampling methodology may be found in U.S. Department of Housing and Urban Development (1978: 377-378).
6. The sample communities' applications for CDBG funds were coded into a detailed format indicating standard HUD budget line items, broad and specific activity groupings, a location code distinguishing residential, nonresidential and citywide activities, and the census tract in which each activity occurs. The application data were matched with the demographic characteristics of each census tract based

upon the 1970 census. It is this matching of application and census data that reveals the outcome of the local CDBG decision process and is consistent with the socio-spatial approach to outcome analysis.

REFERENCES

ARKIN, H. (1963) Handbook of Sampling for Auditing and Accounting. New York: McGraw-Hill.

BARDACH, E. (1977) The Implementation Game: What Happens After a Bill Becomes a Law. Cambridge, MA: MIT Press.

Congressional Quarterly Weekly Report (1978) "House panel challenges Harris move to boost block grant aid to poor." 36 (June 3): 1398-1399.

DOMMEL, P. R. et al. (1978) Decentralizing Community Development. Washington, DC: U.S. Department of Housing and Community Development (June).

EMBRY, R. C. (1977) Management of the Community Development Block Grant Program. Washington, DC: U.S. Department of Housing and Urban Development, Notice CPD 77-10 (April 15).

——— (1978) Review of Entitlement Grant Applications for Fiscal Year 1978. Washington, DC: U.S. Department of Housing and Urban Development, Notice CPD 78-9 (April 28).

FRIEDEN, B. J. and M. KAPLAN (1975) The Politics of Neglect: Urban Aid from Model Cities to Revenue Sharing. Cambridge, MA: MIT Press.

HAAR, C. M. (1975) Between the Idea and the Reality. Boston: Little, Brown.

LIEBSCHUTZ, S. F. (1977) "Community development block grants: who benefits?" Presented at annual meeting of American Political Science Association. Washington, DC, September.

MEEKER, D. O., Jr. (1976a) Testimony Before the Senate Committee on Banking, Housing and Urban Affairs, Oversight Hearing on Community Development Block Grant Program. 94th Cong. 2nd Sess. (April 26).

——— (1976b) Clarification of Review Requirements. Washington, DC: U.S. Department of Housing and Urban Development (May 12).

MOYNIHAN, D. P. (1969) Maximum Feasible Misunderstanding. New York: Free Press.

NATHAN, R. P. et al. (1977a) Block Grants for Community Development. Washington, DC: U.S. Department of Housing and Urban Development.

——— (1977b) "Monitoring the block grant program for community development." Political Science Quarterly 92 (Summer): 219-244.

National Association of Housing and Redevelopment Officials (1976) Community Development Monitoring Study. Washington, DC: Author.

RABINOVITZ, F. et al. (1976) "Guidelines: a plethora of forms, authors, and functions." Policy Sciences 7 (December): 399-416.

ROSENFELD, R. A. (forthcoming) "Local implementation decisions for community development block grants." Public Administration Review.

Southern Regional Council (1976) A Time for Accounting: The Housing and Community Development Act in the South. Atlanta, GA: Author.

SUNDQUIST, J. L. and D. W. DAVIS (1969) Making Federalism Work. Washington, DC: Brookings Institution.

U.S. Congress (1974) 1974 Housing and Community Development Act. Pub. L. 93-383, 88 Stat. 633 (August 22).

——— (1975) House of Representatives, Committee on Banking, Currency, and Housing, Subcommittee on Housing and Community Development. Evolution of Role of the Federal Government in Housing and Community Development. 94th Cong., 1st Sess.
——— (1977a) Housing and Community Development Act of 1977. Pub. L. 95-128, 91 Stat. 1111 (October 12).
——— (1977b) Senate, Committee on Banking, Housing and Urban Affairs. S. Rept. 95-175, 95th Cong., 1st Sess.
——— (1978) Housing and Community Development Amendments of 1978. Pub. L. 95-557, 92 Stat. 2080 (October 31).
U.S. Department of Housing and Urban Development (1975) Community Development Block Grant Program: First Annual Report. Washington, DC: U.S. Government Printing Office.
——— (1977) Community Development Block Grant Program: Second Annual Report. Washington, DC: U.S. Government Printing Office.
——— (1978) Community Development Block Grant Program: Third Annual Report. Washington, DC: U.S. Government Printing Office.
U.S. Federal Register (1975) June 9: 24692-24716.
——— (1977a) February 2: 6504-6505.
——— (1977b) October 25: 56466.
——— (1978a) February 1: 4382.
——— (1978b) March 1: 8450-8474.
U.S. General Accounting Office (1976) Report of the Comptroller General of the United States. Meeting Application and Review Requirements for Block Grants Under Title I of the Housing and Community Development Act of 1974. Washington, DC: U.S. Government Printing Office.
Weekly Compilation of Presidential Documents (1974) 10, 34: 1059-1061. Washington, DC: U.S. Government Printing Office.

10

Urban Development Action Grants:
Design and Implementation

JOHN R. GIST

☐ ONE OF THE NEWEST FEDERAL EFFORTS to revitalize central cities is the Department of Housing and Urban Development's Urban Development Action Grand (UDAG) program, which was passed by Congress on October 12, 1977 as Sec. 119 of the Housing and Community Development Act of 1977. UDAG was the first major urban initiative of the Carter Administration, and presaged by six months the announcement of the administration's urban policy.

The objectives of UDAG are to alleviate physical and economic deterioration by providing assistance "for economic revitalization in communities with population out-migration or a stagnating or declining tax base, and for reclamation of neighborhoods, having excessive housing abandonment or deterioration" (U.S. Federal Register, 1978: 1605).

Essentially, the UDAG program provides capital subsidies to encourage businesses to move into economically distressed cities or to persuade those already located in such cities to remain or expand their facilities and operations. By doing so, it is hoped, the increased investment will create new jobs and help retain existing ones, and will increase the property tax base and begin a process of economic development that will improve the economic and fiscal position of the cities receiving grants.

The UDAG program represents one case of a trend away from the formula entitlement grant programs that were the major thrust of the Nixon Administration's New Federalism. The latter was characterized by broader-purpose "block grants" with formula-based allocation mechanisms, such as General Revenue Sharing, Community Development Block Grants (CDBG) and the Comprehensive Employment and Training Act. Grants under the UDAG program are project based, and funding is discretionary, as under the older categorical programs. Although it is part of the community development title of the 1977 amendments, UDAG is somewhat narrower in focus than CDBG, i.e., its sole purpose is economic development, which is only one of many eligible activities under CDBG. Nevertheless, many of the activities conducted with UDAG funds are conventional community development activities, such as land acquisition, infrastructure improvements, housing rehabilitation, etc., which could be carried out with CDBG funds.

One of the compromises made to assure enactment of the program was the inclusion of a set-aside for cities under 50,000 that are outside Standard Metropolitan Statistical Areas (SMSAs). No less than one-quarter of program funds (originally authorized for three years at $400 million annually with an additional $275 million requested for FY 80) are to be allocated to cities under 50,000 that are not central cities of SMSAs, the remainder going to distressed metropolitan cities and urban counties.

In this chapter, three sets of concerns related to the design and operation of the UDAG program will be considered: (1) a brief comparison of UDAG with two other similar redevelopment efforts—urban renewal and the Title IX program (the Special Economic Development and Adjustment Assistance Program) of the Economic Development Administration (EDA); (2) a discussion of some of the major operational issues surrounding the program; and (3) an analysis of the distribution of Action Grants through the first seven quarters.

OTHER REDEVELOPMENT EFFORTS

Urban economic development has become the latest fad in urban policy circles. The Department of Housing and Urban Development (HUD) has had relatively little experience in the field of economic development. Nevertheless, HUD has been given a leading role in the implementation of federal economic development efforts. The UDAG program was its first major economic development venture. Although there are many similarities between UDAG and the urban renewal program, which was "folded in," along with several other programs, to the CDBG program in 1974, there are also crucial differences.

URBAN RENEWAL

The initial emphasis of urban renewal was on residential development—clearing slums and constructing new housing to replace them. As the program evolved, greater emphasis was placed on commercial and industrial areas of cities, particularly in central business districts. In its commercial phase, urban renewal required that local governments submit plans to the Renewal Assistance Administration for certification. Projects began with the designation of a blighted area as a redevelopment area. A local public agency (either the city or an urban renewal agency) would purchase the land, assemble it, relocate tenants if necessary, clear and prepare the site and make necessary improvements (e.g., water and sewer systems, streets, lighting, etc.). It would then dispose of the land by sale or lease to a private or public developer. Up to two-thirds of the net cost to the local agency would be paid for by the federal government (Rothenberg, 1967).

In terms of many of the activities described above, UDAG is very similar to urban renewal. A typical UDAG project will involve land acquisition, site clearance and preparation, and the provision of infrastructure improvements for a private developer. The land may be sold or leased back to the developer. A sizable number of UDAG projects are located on urban renewal sites. In fact, the availability of vacant urban renewal sites points up perhaps the major difference between UDAG and urban renewal. Many urban renewal projects faltered because anticipated investment by a private developer did not occur. Projects that were initiated by cities were not completed for want of private capital, and sites planned for renewal remained vacant.

The UDAG program's critical difference from urban renewal is that the private capital component of a proposed UDAG project must be committed *in advance* and local governments must show evidence that the private sector investment is firmly committed. The evidence of the firmness of that commitment is to be provided by applicants in the form of letters of intent from investors/developers or financial institutions stating clearly what the nature of the investment is, the relationship to the requested UDAG funding, details of the investment package, amounts, types, terms, and conditions of loans, and indications of a willingness to sign legally binding commitments (U.S. Department of Housing and Urban Development, n.d.). Eventually, before any Action Grant funds can be legally obligated, the legally binding commitments from all project participants must be accepted by HUD. It is anticipated that these requirements will make it more certain that an approved project will succeed, i.e., it will be completed within a reasonable period of time. The requirements do not preclude the possibility of failure—HUD has already withdrawn awards from a dozen cities due to loss of private commitment—but they are expected to reduce the likelihood of these failures. However, the strong emphasis on legally binding commitments raises certain issues, such as whether UDAG funds are truly stimulating private investment and whether the emphasis on selecting projects that will succeed has altered the intent of the program. These issues will be discussed in a later section.

Other differences between UDAG and urban renewal include targeting on distressed cities, the lack of a restriction on the federal share of the UDAG grant, and the emphasis on job creation and other economic impacts as selection criteria.

EDA TITLE IX

EDA's Title IX program bears some similarity to UDAG in that it is targeted at communities that are economically distressed. But as the EDA program's guidelines state, the funds were to be allocated to areas suffering "acute employment dislocation as a result of a localized problem in contrast to areas of long-term unemployment and/or low family income which are targets of EDA's regular programs" (U.S. Department of Commerce, 1977: 52). UDAG is targeted at cities experiencing long-term decline rather than short-term or cyclical dislocations, and distress is measured in a more comprehensive way than for Title IX. Title IX is not targeted at central city areas.

Another major difference is that UDAG places tremendous emphasis on leveraging private sector investment, whereas the guidelines of Title IX are concerned more with leveraging funds from other federal entities. Thus, there is no emphasis placed on the "public/private partnerships" that the national urban policy advocates.

Third, the small size of the Title IX program (outlays in FY 77 were approximately $77 million) will prevent it from having a major impact on central city revitalization.

POLICY ISSUES

Although it is too early to evaluate UDAG or to foresee all the problems that might befall the program in the future, a number of issues have already surfaced that deserve attention. (In that connection, also see the chapters by Ross, and by Cho and Puryear in this volume.)

TARGETING

In a time of scarcity, targeting has become a critical element in the design of federal programs. Several federal grant programs target aid to jurisdictions by means of various allocation formulas, and the formulas are evaluated in part by how accurately they target aid to the desired places.

UDAG is supposed to target funds to cities or urban counties that are physically and economically distressed, distress being measured by scores on certain demographic characteristics. However, grants are discretionary, and a determination that a city is distressed only makes it *eligible* for UDAG. It does not thereby entitle it to any funds.

The UDAG eligibility criteria for metropolitan and small cities differ slightly. The program regulations identify minimum levels of distress for metropolitan cities on six different measures. The minimum standards are the medians for all metropolitan cities. The measures and associated minimum thresholds as of January 1979 were:

(1) percentage of housing units constructed prior to 1940 (34,15%);
(2) net increase in per capita income, 1969-1974 ($1,424 or less);
(3) percentage rate of population growth from 1960 to 1975 (15.52% or less);
(4) average yearly rate of unemployment for 1978 (6.98% or more);
(5) rate of growth in retail and manufacturing employment 1967-1972 (7.08% or less); and
(6) percentage poverty (11.24% or more).

Metropolitan city and urban county applicants must generally qualify on at least three of these six minimum standards.[1] In addition, they must be able to demonstrate achievement of results in providing low-income housing and equal opportunity. The criteria for small cities, although similar, are fewer in number because reliable data on certain of the measures are not available for small cities.[2] On the basis of these factors and thresholds, there are 323 metropolitan cities eligible for Action Grants and over 2,000 eligible small cities.

These *eligibility* criteria are expected to guarantee that Action Grants will go only to distressed cities. But they are minimum criteria. Obviously, among the 300-odd metropolitan cities and over 2,000 small cities, some are more distressed than others. Thus, the eligibility criteria alone provide only a weak targeting mechanism because it is possible that Action Grants could go to cities that satisfy only the minimum number of eligibility standards.

This weakness in eligibility standards might be compensated for, however, by the UDAG *selection* criteria, which also include indicators of distress. The Housing and Community Development amendments of 1977 incorporate a formula, commonly known as the "impaction" formula, as the primary criterion in *selecting* projects for Action Grant awards. The 1977 legislation added this formula to the original CDBG formula (which included population, overcrowded housing, and poverty weighted twice) to improve the targeting of CDBG funds to distressed cities. "Impaction" is a composite of age of housing, poverty, and growth lag. These are defined more precisely in the UDAG regulations which state that

> HUD shall select from such feasible and effective proposals those to be funded on the following bases:
>
> (1) as the primary criterion, the comparative degree of physical and economic distress among applicants, as measured by the differences in the factors set forth below, which shall be assigned the relative weights as follows:
>
> (i) The percentage of their total housing stock that was built prior to 1940: .5;
>
> (ii) The percentage of their total current population that was in poverty in 1970: .3; and
>
> (iii) The degree to which their population growth rate lags behind that of all metropolitan cities: .2 [U.S. Federal Register, 1978: 1607].

The priority given to distress in the selection criteria would suggest a highly targeted selection process. However, the key words in the section of the regulations quoted above are "feasible and effective." In the implementation of the Action Grant program, those words have come to mean "having strong private financial backing." Thus, the authorizing legislation and program regulations have been interpreted in such a way as to give priority to private investment over distress as a selection criterion. The extent and firmness of private financial commitment was, indeed, intended to be a selection factor in the program, but only *after* the degree of distress, as measured by "impaction," was considered. The operation of the selection process, therefore, diverges in practice from the prescriptions of the regulations.

This divergence of the selection process from that prescribed in regulations does not necessarily preclude the possibility that grants are being awarded to the most distressed cities. As Table 10.2 below shows, two-thirds of the projects have been awarded to cities in the top 40% of all eligible cities ranked by level of distress. Despite a conservative bias in the selection process, HUD seems to have fulfilled the spirit of the law, if not the letter in funding UDAG projects.

However, the priority given to private investment in the selection process may still bias the selection of projects in a way that favors low-risk projects and disadvantages higher risk ones. Even distressed cities such as New York, Baltimore, and Washington, DC, present investment opportunities to the private sector that are highly profitable and relatively free of risk. There is an understandable reluctance on the part of HUD officials to risk failure.

LEVERAGING

"Leveraging" is a term popularized in government circles which denotes the attraction of additional private or public funds by an initial infusion of federal funds. Without substantial private sector financial commitment to a project, there is no chance of a city obtaining an Action Grant. That requirement, in particular, is what differentiates UDAG from earlier federal redevelopment programs.

In their grant applications, cities provide estimates of the private investment leveraged and the number of jobs that are expected to be created by a project. These estimates are evaluated in area offices by the Economic and Marketing Analysis Divisions of HUD, but are usually not altered. When project proposals reach the HUD Central Office, the reliability of the job estimates is not questioned so much as is the firmness of the private investment that creates them. If any part of a project looks "soft," in the sense that the private investment is not firmly committed, the portion of the total jobs that is attributable to that part of the project will be excluded from estimated benefits.

Despite the requirement for private sector investment, there is, nevertheless, a real possibility that a substitution or displacement effect will occur, i.e., that the public or private investment will merely replace investment that would have occurred anyway. Any long-term evaluation of the Action Grant program must consider the extent to which such substitution occurs. One type of substitution that may occur is financial, where public (UDAG) dollars are substituted for private investment.

On this issue of public-for-private substituion, there seems to be a potential contradiction built into the concept of the program. The UDAG program places a premium on the leveraging of private funds. Yet, from the congressional testimony given prior to the enactment of the program, a basic justification for the program was that UDAG would provide the last bit of funding necessary to complete the packaging of a project and put it into action. If the latter is truly the case, then the possibility exists that UDAG funds are not exercising leverage at all, but simply adding on to investment already planned. A recent U.S. General Accounting Office (1979) study of eighteen UDAG projects claims that in some cases the UDAG funds have been applied for only *after* the private firms have already committed themselves. These cases are extreme, and perhaps rare, but they suggest that the claims made for leveraging should be treated with a healthy skepticism.

Another type of substitution that can occur is geographic substituion, which can take on several forms: on-site displacement, intrajurisdictional substitution, intrametropolitan substitution, or intermetropolitan substitution.[3] On-site displacement occurs when one firm replaces another, whether in the same industry or not. In cases of intrajurisdictional, intrametropolitan, or intermetropolitan substitution, a firm relocates or is replaced by another firm in the same industry or serving the same market. HUD regulations prohibit the use of UDAG funds to facilitate the relocation of firms from one "area" to another. But this prohibition does not apply to moves *within* a metropolitan area. Thus, the fourth type of geographic substitution is not permitted under UDAG, while the others are (U.S. Federal Register, 1978: 1606).

The restriction on intermetropolitan relocation is no doubt motivated by a desire on the part of HUD officials to discourage the movement of commercial and industrial employers, and consequently the movement of jobs, from "snowbelt" to "sunbelt" cities.

What is less understandable is why movements within a metropolitan area are considered acceptable, since this could still be an interjurisdictional move. That is, a business wanting to move from the central city to a neighboring suburb could not be denied assistance for this reason. The rationale may be that interjurisdictional movements that remain within a metropolitan area will not deprive people of jobs (assuming the new location is within commuting distance), whereas intermetropolitan relocation would cause jobs to be lost. However, the relocation from a central city to a neighboring suburb may have serious negative consequences for the tax base of the central city, an outcome that HUD would surely not favor but which the regulations, nevertheless, do not prohibit.

PROJECT TYPES

The Housing and Community Development Act of 1977 and the UDAG regulations both specify that UDAG funds are to be allocated in a "reasonably balanced" manner among projects that restore deteriorated neighborhoods, reclaim for industrial purposes underutilized real property, and renew commercial employment centers (U.S. Federal Register, 1978: 1607). This requirement almost immediately became a source of controversy in the program. In the first round of Action Grant awards, which were announced in early April of 1978, nearly one-third of the fifty projects funded were hotel or hotel/convention center complexes. Very few neighborhood projects were selected. There was an immediate outcry from neighborhood groups and criticism that HUD was unresponsive to the needs of low- and moderate-income people. HUD Secretary Patricia

Harris defended the hotel projects, saying that they employed many low-income people and that it was possible for these employees to move into better jobs eventually, e.g., they might start out as dishwashers and end up as chefs (Washington Post, 1978). As a result of this uproar and the concern expressed by Senator Proxmire, who chairs the appropriations subcommittee on the HUD budget, a premium was placed on selecting neighborhood projects in the second round of awards.

One of the problems with the existing nominal categories of industrial, commercial, and neighborhood projects is that they are not mutually exclusive. While industrial and commercial projects are defined by use, neighborhood projects are defined by location. Thus, it is possible to have industrial or commercial projects located in neighborhoods. Projects having a mixture of purposes, such as commercial space and housing, are not uncommon either. Thus, there may be differences of opinion in project categorization. Since no clear guidelines on the categorization of projects exist, there is also room for manipulation. At the end of a reviewing period, when all projects have been selected but not yet announced, last-minute changes have been made in project classification in order to achieve better balance among projects. Thus, for example, shopping malls located in neighborhoods have been shifted from the larger commercial to the underrepresented neighborhood category.

The greatest ambiguity exists with respect to the neighborhood category. A more meaningful category than "neighborhood" would be residential/housing which would include any project where the major emphasis (i.e., largest expenditure) is on housing construction or rehabilitation. This category has the advantage of being defined by use, and would reflect the fact that the majority of neighborhood projects involve housing anyway. Ambiguities would still exist in the case of projects that have both commercial and housing components. In such cases, the mixed nature of the projects should be acknowledged by creating a separate category for mixed projects.

DISTRIBUTION OF UDAG PROJECTS AND FUNDS

This section analyzes the distribution of UDAG projects, funds, and private investment among metropolitan and small cities, among types of projects, levels of distress, and regions. The data in the following tables include all Action Grant projects funded through the first quarter of FY 79. This includes four rounds of metro city and only three rounds of small city projects, because the small city program got underway several months after the metro city program.

METROPOLITAN VS. SMALL CITIES

The UDAG legislation requires that one-quarter of program funds are to go to "small cities," which are defined in regulations as "cities under 50,000 which are not central cities of an SMSA" (U.S. Federal Register, 1978: 1605). A total of 309 Action Grants were awarded through the first quarter of FY 79. However, the tables show only 304 projects, because five metropolitan city awards which have been rescinded by HUD were excluded.[4] The distribution of these projects, UDAG funds, and private investment among metropolitan and small cities are reported in Table 10.1.

It is evident that metropolitan Action Grant awards are four times as large, on average, as the small city awards. The small city program has thus far been characterized by a large number of small grants. However, these smaller grant amounts for the small cities have attracted higher rates of private investment than metropolitan city projects. The latter leverage three dollars less per UDAG dollar, on average, than do small city projects. The large amounts of private funds attracted to small city projects may reflect the relatively greater attractiveness of nonmetropolitan sites due to lower labor costs and lower tax rates, as well as the availability of cheaper land. This would particularly be true for industrial parks which are very common among the small city projects. There are relatively few industrial projects among the metropolitan cities, as we shall see below.

TABLE 10.1   Distribution of UDAG and Private Funds by Project Type (dollar amounts in thousands)

Type of Project	Number of Projects	UDAG Funds Awarded	Private Investment	Private $/ UDAG$	Average Award
Industrial	112	135,859	906,321	6.67	1,213
Metro	34	76,956	338,841	4.42	2,263
Small	78	58,903	567,480	9.63	955
Commercial	112	262,079	1,667,074	6.36	2,340
Metro	65	229,702	1,438,079	6.26	3,534
Small	47	32,377	228,995	7.07	689
Neighborhood	80	185,562	882,400	4.76	2,320
Metro	59	163,857	745,618	4.55	2,777
Small	21	21,705	136,782	6.30	1,034
Total	304	583,500	3,455,795	5.92	1,919
Metro	158	470,515	2,522,538	5.36	2,978
Small	146	112,985	933,257	8.26	774

TYPE OF PROJECT

As noted earlier, the UDAG regulations prescribe that a "reasonable balance" be maintained among neighborhood, industrial, and commercial projects. The shortcomings of this classification scheme have already been noted. Table 10.1 shows the metro and small city project distribution by type, through seven quarters. For the purposes of the display in the table, the classification of projects used by HUD is not altered. The figures are taken from the UDAG Project Selection Data File and from assorted HUD news releases.

The criticism from neighborhood groups regarding the overemphasis on commercial projects, particularly hotel projects, was noted earlier. While the distribution of projects became somewhat more balanced in subsequent quarters, the commercial category still receives 45% of all UDAG funds awarded, due primarily to the extraordinarily large awards given to commercial projects in metropolitan cities. Commercial projects receive not only the highest total allocations, but the largest average UDAG awards as well. Commercial uses enjoy a comparative advantage over industrial uses in central city locations, explaining the predominance of commercial projects in the metro city category. They have the highest leveraging ratios of the metropolitan projects, attracting approximately 40% more private investment than do industrial or neighborhood projects.

Table 10.1 reveals that, in aggregate, small cities seem to have greater leveraging potential than metropolitan cities. This pattern also holds for the different types of projects. The greatest advantage enjoyed by non-metropolitan cities is in the industrial category, where small city projects are more than twice as effective in attracting private investment as are metro city projects. Again, given the relative attractiveness of industrial uses in nonmetropolitan areas, it is not surprising that over 50% of the 146 small city awards have gone to industrial projects.

LEVEL OF DISTRESS

The UDAG program regulations state that, after the feasibility and effectiveness of a project have been determined, the primary selection criterion shall be the "comparative degree of physical and economic distress among applicants," which is to be measured by the relative ranking of applicants on the impaction formula discussed earlier. This impaction ranking does seem to play a role in the process of review and in the deliberations prior to the allocation of UDAG awards.

Evidence that levels of distress have been important factors in the selection process was reported by HUD in a preliminary study of the

TABLE 10.2   Distribution of UDAG and Private Funds by Distress
Quintiles for Metropolitan City Projects (dollar
amounts in thousands)

Distress Quintile	Number of Projects	UDAG Funds Awarded	Private Investment	Private $/ UDAG$	Average Award
1	51	123,954	549,191	4.43	2,430
2	50	166,343	997,045	5.90	3,327
3	27	106,167	554,688	5.22	3,932
4	16	47,586	308,480	6.48	2,974
5	14	26,937	136,149	5.05	1,924
Total	158	470,987	2,545,553	5.40	2,978

UDAG program. In that study, which included five rounds of projects
funded through December of 1978, HUD found consistently higher levels
of distress on five of six eligibility factors for those cities receiving grants
compared to nonrecipients. The sole exception to this pattern was the
poverty factor, on which the nonrecipients consistently registered greater
levels of distress (U.S. Department of Housing and Urban Development,
1978: 12-15). This adds empirical evidence to the conclusion that the
measures of distress have played an important role in project selection.
Table 10.2, which looks at only metropolitan city projects, adds more
support to that conclusion. It allocates all funded projects among five
"distress quintiles," the first quintile representing the 20% of all eligible
metropolitan cities that are the most distressed (based on impaction
ranking), the second quintile being the next most distressed, and so on.
Table 10.2 shows that nearly two-thirds of all projects were awarded to
cities that are among the most distressed 40% of *all eligible* metropolitan
cities.[5] It also shows that the number of grants awarded for each quintile
declines as we go from the most distressed group to the least distressed.
However, the largest average grant went to cities in the middle quintile,
and the most distressed quintile received only the fourth largest average
grant.

Other things being equal, we would expect the degree of private sector
support to be lower in more distressed cities. However, from evidence in
Table 10.2 there seems to be no systematic relationship between levels of
distress and ability to attract private investment. While the most distressed
quintile of cities has the lowest leveraging ratio, the next most distressed
quintile has a higher average leveraging ratio than either the third or fifth
quintiles.

REGION

A fundamental programmatic objective of UDAG is to target economic development assistance on distressed cities, wherever they happen to be located. While it was expected that some Southern and Western cities would qualify under the UDAG eligibility criteria, there was also a clear expectation that the majority of eligible cities would be in the Northeast and North Central regions. Thus UDAG has been embroiled in the controversy between "sunbelt" and "snowbelt" cities over the allocation of federal funds.[6] Table 10.3 presents the distribution of Action Grant funds and associated private investment by region.[7] Not surprisingly, it shows that the Northeast and North Central regions have done very well in obtaining Action Grants. The two regions combined have received two-thirds of all UDAG grants awarded and a like percentage of all UDAG funds awarded. They have also attracted two-thirds of the private investment that has been leveraged thus far. The high percentage of private investment for these combined regions is due in part to the North Central

TABLE 10.3   Distribution of UDAG and Private Funds by Region (dollar amounts in thousands)

Region	Number of Projects	UDAG Funds Awarded	Private Investment	Private $/ UDAG$	Average Award
Northeast	125	245,067	1,226,844	5.01	1,961
Metro	75	190,394	796,755	4.18	2,539
Small	50	54,673	419,455	7.67	1,093
North Central	77	153,787	1,082,318	7.04	1,997
Metro	41	131,173	932,820	7.11	3,199
Small	36	22,614	149,498	6.61	628
South	54	78,318	543,251	6.94	1,450
Metro	19	58,132	266,106	4.58	3,060
Small	35	20,186	277,145	13.73	577
Midwest	24	31,678	290,034	9.16	1,320
Metro	5	21,993	244,730	11.13	4,399
Small	19	9,685	45,304	4.68	509
Far West	24	74,708	341,856	4.58	3,113
Metro	18	68,823	305,142	4.43	3,824
Small	6	5,885	36,714	6.24	981
Total	304	583,558	3,484,303	5.97	1,920
Metro	158	470,515	2,545,553	5.40	2,978
Small	146	113,043	928,116	8.21	774

region's high leveraging ratio—it attracts two dollars more per UDAG dollar than do projects in the Northeast. The South ranks third among the regions with 18% of all grants and 13% of all funds awarded.

OVERVIEW

Although HUD has been awarding Action Grants since April of 1978 and some projects have been completed, the vast majority of those funded are either in their early stages or not as yet underway. The UDAG program office achieved remarkably swift results in the period immediately after the enactment of the program in getting the program established. Regulations were written and accepted, the program handbook was approved and the first set of projects were selected and announced within six months after enactment. All this was done with a very small staff. However, the performance in project implementation—with a substantially increased staff—has been far less satisfactory. In fact, the delay in project starts has been a source of dissatisfaction even within HUD. Delays have been caused by a variety of factors, including environmental, equal opportunity, and other clearances. But the two most significant causes of delay have been the slow contract negotiation process[8] and the requirement of HUD acceptance of the legally binding private commitment.

Because of the lack of data on completed projects, any attempt to evaluate the UDAG program's impacts as of this writing would necessarily be premature. I have attempted in this chapter the more modest task of elucidating some of the major issues that have arisen in the early stages of the program and demonstrating how UDAG funds have been distributed in terms of key programmatic factors. Given the limited scope of the task, a few tentative conclusions emerge.

First, the UDAG program seems to differ little from urban renewal except for its emphasis on having private sector investment firmly committed to the project in advance of HUD funding. This assertion is based on the fact that the project expenditures as reported in applicants' budgets are allocated for the same kinds of activities undertaken with urban renewal.

Second, the intention of Congress that level of distress be the primary selection criterion has been circumvented in the implementation of the program by using firm private commitment as a prior criterion. However, this does not seem to have worked a hardship on distressed metropolitan cities since two-thirds of the metropolitan grants awarded went to cities among the most distressed 40% of all those eligible. Thus, it appears that the selection process has worked relatively well in targeting severely distressed cities.

Third, the concept of the program as originally advertised to Congress seems inconsistent with the assertion that UDAG successfully leverages private investment. It is at least problematic, although very difficult (if not impossible) to demonstrate, that much of the private sector investment "leveraged" by UDAG would have occurred anyway.

Finally, it is difficult to determine whether the program has fulfilled the legislative requirement that a "reasonable balance" be maintained among project types, because no definition of "reasonable balance" was ever given. However, a concern that this "balance" has not always been achieved is evidenced by the shuffling of projects (noted earlier) from the commercial into the neighborhood category.

NOTES

1. However, if the applicant's percentage of poverty is less than one-half of the median for all metropolitan cities, then four standards must be met. If an applicant's percentage of poverty is more than one and one-half of the median for all metropolitan cities and its absolute per capita income is below the median, a unique distress factor may also be considered.

2. Cities of less than 2,500 are generally ineligible unless they show a capacity to carry out a comprehensive community development program, demonstrate that their distress can only be resolved through a UDAG, and meet three minimum standards of distress: age of housing, per capita income, and poverty. Cities between 2,500 and 10,000 must meet all three of the distress standards cited above. Cities of 10,000 to 25,000 must meet three of four minimum standards—the three above plus the population decline standard (standard no. 3 above). Cities of 25,000 to 50,000 must meet three of five standards—the above standards plus the job lag standard (standard no. 5 above) where data are available.

3. This discussion of substitution is based on Jacobs and Roistacher (1979).

4. As of this writing, a total of seven small city projects have also been withdrawn but too late to be reflected in these tables.

5. The data in Tables 10.2 and 10.3 differ slightly in some respects from Table 10.1 because they are based on different data sources. This was necessary because HUD has been updating its data base on all active projects, so that some of the data have changed slightly over time.

6. HUD has recently issued a report on the question of "pockets of poverty" and whether cities that have such areas of distress should qualify for the UDAG program (U.S. Department of Housing and Urban Development, 1979). The study was a response to the "Tower amendment," which would make cities with distressed *areas* that meet UDAG *citywide* criteria eligible for Action Grants. HUD has supported a plan to channel up to 15% of UDAG funds to these areas. This dilution of the targeting emphasis of the program is a concession made by HUD to facilitate passage of an added $275 million for the UDAG program in the FY 80 budget.

7. The five regions shown here are composites of the ten HUD administrative regions. The Northeast is comprised of the New England states, New Jersey, New York, Pennsylvania, Maryland, Delaware, Virginia, West Virginia, and the District of

Columbia. The North Central region includes Ohio, Michigan, Indiana, Illinois, Wisconsin, and Minnesota. The South includes North Carolina, South Carolina, Georgia, Florida, Alabama, Mississippi, Tennessee, Kentucky, Louisiana, Arkansas, Texas, Oklahoma, and New Mexico. The Midwest includes Iowa, Missouri, Kansas, Nebraska, North Dakota, South Dakota, Montana, Colorado, Utah, and Wyoming. The Far West includes Arizona, Nevada, Idaho, California, Oregon, Washington, Hawaii, and Alaska.

    8. Until April of 1979, UDAG had one staff member negotiating all the contracts with Action Grant recipients. At that time, the staff members responsible for reviewing and evaluating projects were also given responsibility for writing grant contracts, a step intended to reduce the lag in project starts.

REFERENCES

Congressional Quarterly (1978) Urban America: Policies and Problems. Washington, DC: Author (August).

JACOBS, S. and E. A. ROISTACHER (1979) "The urban impacts of HUD's urban development action grant program, or where's the action in action grants?" Presented at a conference on the urban impacts of federal policies, Washington, DC, February 8-9.

ROTHENBERG, J. (1967) Economic Evaluation of Urban Renewal. Washington, DC: Brookings Institution.

U.S. Department of Commerce (1977) An Administrative Evaluation of the Special Economic Development and Adjustment Assistance Program. Washington, DC: Author.

U.S. Department of Housing and Urban Development (n.d.) Urban Development Action Grant Program: Guidance on Letters of Intent. Washington, DC: Author.

––– , Office of Community Planning and Development (1978) Preliminary Review of the Urban Development Action Grant Program. Washington, DC: Author.

––– (1979) Pockets of Poverty: An Examination of Needs and Options. Washington, DC: Author.

U.S. Federal Register (1978) January 10: 1602-1610.

U.S. General Accounting Office (1979) "Improvements needed in selecting and processing urban development action grants." March 30.

Washington Post (1978) "U.S. giving cities $150 million for redevelopment projects." April 17.

PART IV

A COMPARATIVE PERSPECTIVE

11

Neighborhood Revitalization:
The British Experience

H.W.E. DAVIES

□ CONSIDERABLE PROGRESS HAS BEEN MADE since World War II in improving the quality and condition of housing in Britain.[1] The recent *Housing Policy Review*[2] estimated that 10 million households in 1951 were living in dwellings that in some respect were below standard.[3] By 1976, the number had been brought down to 2.7 million when the three chief problems were unfitness (affecting 5% of the total housing stock); lacking exclusive use of one or more basic amenities (11%); and poor repair (13%). Many dwellings, especially those built before 1919, were affected by all three problems.

Progress during the 1950s and 1960s relied almost entirely on slum clearance, but emphasis has shifted in the last ten years so that now the chief method is the improvement of older houses including their modernization to meet current standards, including conversion into dwellings of more appropriate size and repair.

Neighborhood revitalization, or area improvement as it is more generally known in Britain, has contributed to this progress. At first, it was seen as an expedient method of maintaining substandard houses in a tolerable condition until they could be demolished under slum clearance programs. But area improvement increasingly has become an end in itself for revitalizing older neighborhoods, thus preventing the necessity of their clearance and demolition.

This chapter traces the reasons for the change in policy from clearance to improvement and for the changing emphasis in the improvement policy itself from temporary expedient to revitalization. It assesses the contribution of area improvement programs to national housing objectives and contrasts this with their role in local urban planning objectives. But clearance and improvement can properly be understood only in the context of the wider spectrum of housing policies. Accordingly, the chapter opens with a brief description of the housing situation in Britain.

HOUSING IN BRITAIN

Housing has been a subject of government concern in Britain for well over a century. Government policies have widened to cover virtually every aspect of housing. At first the aim was to regulate standards of fitness and enforce building regulations; later, it came to include the clearance of slums, the control of rents, and the provision of new housing by the public sector, initially for the "working classes" but later to meet more general needs not catered to by the private market. During the last half century, government has provided incentives for "owner occupation" (i.e., home ownership) and, most recently, for the improvement of the older housing stock. Throughout, the pattern has been for national government to enact legislation and allocate resources for housing policies. Local authorities then have had considerable discretion in the detailed implementation of those policies.

The story is long and complex and has resulted in a very distinctive national housing stock compared with America. Briefly, there are four principal types of housing tenure: owner occupation, privately rented accommodation, council tenancies, and housing associations.

Owner occupation is mainly in new houses built for sale by construction companies but also by the purchase of older private rented houses. Housing policies have encouraged this form of tenure by giving tax relief on the interest payable on mortgages, through exemption from capital gains tax or any tax on the imputed rent arising from property, and from special assistance for so called "first time buyers." It has become a rapidly expanding share of the total stock, rising from 31% in 1951 to 55% in 1976.

Privately rented accommodation was built mainly before 1919 and is let either unfurnished or furnished. Most landlords own very few properties; only a small number are owned as commercial investments by property companies. Most privately rented accommodation is subject to rent control, and tenants have considerable security of tenure, especially where the landlord is not resident. Housing policies, in effect, have so discouraged this form of tenure that there has been virtually no new building since World War II. Rent control, security of tenure, and the lack of tax concessions make it an unprofitable investment for landlords. For tenants, there are no tax reliefs and no prospect of capital gains, though the poorest tenants may be eligible for rent allowances.

Many rented houses were substandard or poorly maintained and have been demolished in slum clearance programs. Many of those in better condition have been bought for owner occupation. As a result of this loss of older private rented housing and the lack of any new construction its share of the total stock has fallen from 52% in 1951 to 15% in 1976.

Council tenancies have been provided by local authorities in general since 1919. They are mainly in newly constructed houses or flats but include some older houses acquired for slum clearance or other purposes and used temporarily as "short life" accommodation pending their demolition. Councils give first priority in the allocation of tenancies to families displaced by slum clearance, who have a statutory right to be rehoused; and next to householders applying through the local authority's waiting list, in which preference is usually given to families who have been resident in the local authority area for a long while and have young children, or are living in unsatisfactory conditions. Central government provides a subsidy to local authorities to assist in the construction of council housing, and council tenancies are let at subsidized rents, the amount varying with each local authority. In addition, rent and rate rebates are available for poor tenants and special provision may be made for the elderly, disabled, or homeless. This has been an expanding sector of the housing stock, from 17% in 1951 to 30% in 1976.

Housing associations form a very small, but new and expanding, form of tenure. Briefly, they are nonprofit-making voluntary organizations which may be set up by any group of citizens, residents, or voluntary associations. They range in size from a single house converted into flats to more than 1,000 dwellings. Housing associations have one of three main objectives: (1) to meet the special housing needs of people excluded from the two dominant tenures; (2) to acquire and improve, or convert, older houses often in some form of multiple occupation; (3) to encourage cooperative ownership by residents. Housing associations receive financial assistance from the Housing Corporation set up by central government to promote this form of tenure. They covered about 1% of the total stock in 1976.

Slum clearance and housing improvement policies are aimed chiefly at housing built before 1919, which comprised just under one-third of the total stock in 1976. But the age composition of the different tenures varied considerably. Only 30% of owner-occupied dwellings and 4% of council tenancies were built before 1919, but no less than 69% of privately rented accommodation. Thus coping with the problems of the older

housing, and the success of area improvement as a policy, is closely bound
up with the future of privately rented housing.

SLUM CLEARANCE AND REDEVELOPMENT

The concept of clearance areas, in which whole blocks of unfit dwellings could be compulsorily acquired by local authorities and demolished,
was introduced in the 1930 Housing Act when a massive slum clearance
program was started. The clearance areas were located in the inner areas of
many large cities and the slums were replaced by four and five storey
walkup blocks of tenements, with the surplus population rehoused in new
council housing estates of ordinary houses on the outskirts of the cities.

The program was halted during the war and resumed only in 1954.
Government asked local authorities to prepare five-year slum clearance
programs, but at first the response was poor. Then, in the 1960s, local
authorities realized that slums were being cleared too slowly even to keep
pace with the spread of obsolescence. Clearance rates were accelerated and
soon overtook those of the 1930s. Nearly 1.4 million dwellings were
demolished between 1951 and 1976.

The clearance program had an essential counterpart in the redevelopment policy: indeed, the two were linked in many cities in a single,
computer program. Redevelopment was carried out by local authorities
who replaced the slums by council housing and also made use of the
cleared land to provide open spaces, extensions for schools, small neighborhood shopping centers and community facilities.

Exclusion of private developers, other than where the land was used for
extensions to the central business district, had several causes. Political
ideology played its part, particularly in the inner areas of some large cities,
but the practical problems were paramount. On the one hand, local
authorities had an overriding duty to rehouse people displaced by clearance. On the other hand, many obstacles prevented private redevelopment.
The private sector lacked powers of compulsory purchase to assemble
compact sites suitable for redevelopment. Clearance areas and their environment were thought to be unattractive for new house building for sale.
But the fundamental problem lay in land values. The market value of a
cleared, inner city site for new housing fell far short of the costs incurred
in its acquisition, clearance, and site preparation (U.K. National Economic
Development Office, 1971). This was sufficient to keep out the private
sector until the first moves were made in 1976 by the City of Liverpool to
invite tenders from private builders for the construction of new houses for

sale on cleared sites, the cost of writing down the land values being borne by the local authority and indirectly by central government. This first example proved successful and is now being repeated in several cities (Fleetwood, 1978).

At first the scale of redevelopment was very similar to the prewar pattern with low-rise walkup blocks of flats, but redevelopment accelerated during the 1960s to match the expanded clearance programs. Increasingly, it took the form of high-rise blocks of flats using industrialized building techniques. Densities were increased to reduce the necessity for relocating the surplus displaced population in suburban council housing estates and to give more opportunity for meeting peoples' preference to be rehoused locally. Industrialized techniques were introduced to increase the rate of construction and obtain economies of scale. High densities were given an impetus by additional subsidies from the central government. But the industrialized building techniques needed large cleared sites. Accordingly, ever larger areas were compulsorily acquired, cleared, and redeveloped by local authorities. Inevitably, such areas contained a wide variety of land uses and housing conditions, including many residential and commercial buildings that were fit and had a useful life, simply to provide land for "efficient and satisfactory [comprehensive] redevelopment." One such area in Liverpool, for instance, covered 31 acres, and involved the compulsory acquisition and clearance of 1,000 unfit dwellings, a further 80 fit dwellings, 150 local shops, public houses, workshops and service industries and two churches, thereby increasing the total cost of acquisition by more than three times compared with the cost of acquiring the unfit dwellings themselves (Gibbons, 1977).

Another feature of the clearance and redevelopment process was the length of time between the first "referral" of an area by a public health inspector for consideration for clearance and the actual completion of redevelopment. The entire process could take ten years or more as policies changed direction, as allowance was made for the statutory rights of objection and public inquiry, and as displaced residents were given several choices for rehousing.

The period up to the actual clearance could blight an entire neighborhood with uncertainty, progressive decay, and dereliction, thereby accentuating the features which made the clearance necessary in the first place (Dennis, 1972; Mason, 1977). The period between clearance and redevelopment could be just as long, particularly if the replacement uses were not for council housing (for which government finance was readily available). If they were for open spaces, school extensions or road schemes (for which finance was in short supply) or for private industrial or commercial development (for which demand was lacking) the cleared areas could remain unused for years, an eyesore and source of blight to adjacent neighborhoods (Burrow, 1978).

Arguments against slum clearance and comprehensive redevelopment mounted during the 1960s even as the program itself gained momentum. The social disruption, the unnecessary destruction of economic enterprise, the actual form of redevelopment, the dispersal to distant council housing estates were all under attack. And the attack was reinforced by the growing awareness of an alternative. The 1949 Housing Act had introduced the idea of *improvement grants* payable by local authorities to the owners of houses (owner-occupiers and landlords) who undertook to improve their property. The 1949 grant was discretionary: owners could apply for a grant equal to 50% of eligible expenditure on improving (but not repairing, or replacing outworn parts of) their houses and it was then up to the local authority whether to award a grant. The intention was to make good intrinsic deficiencies in the design of the house, because of its age, rather than make up for the lack of repair. Thus improvements could include providing basic amenities such as a bathroom, or modern construction devices if none had originally been provided, or converting a large house into self-contained flats.

The cost of the discretionary grants to local authority and homeowner alike was considerable. So the concept was widened in the 1959 Housing Act to include smaller *standard grants* covering 50% of the cost of installing only those basic amenities which were lacking in smaller houses. These were granted to applicants as a right by the local authority. As a result, the numbers of improved dwellings rapidly increased and by 1963 exceeded the number of slum dwellings being cleared each year.

Initially, improvement was seen simply as a means of prolonging the life of dwellings which nevertheless were unfit. The aim was to provide tolerable living conditions for the last years of their lives. Local authorities experimented with different techniques of improvement, including the first halting steps in neighborhood revitalization, by concentrating attention on small areas. Minor improvements were made to the external environment of houses, for instance by repairing back alleyways or landscaping wasteland. Special campaigns were mounted to increase the "take-up" (or use made) of improvement grants by owner occupiers and landlords. But at this stage local authorities lacked the powers and resources to undertake effective area improvement programs (Roberts, 1976).

The attitudes of central government and local authorities to the concept of housing improvement began to change as the social and economic consequences of slum clearance became clearer. Economists devised techniques for comparing the costs and benefits of clearance and redevelopment with those of improvement (Needleman, 1965). One early land use/transportation study built a comprehensive decision-making model for an urban region, identifying areas for clearance or improvement taking into account environmental quality, the costs of total replacement or simple improvement of dwellings and their environment, and the benefits measured by the present worth of the future accommodation thus provided (U.K. Department of the Environment, 1971). Debate focused on questions about rates of interest, the assumed life of property, new or improved, and the comparative quality of the accommodation provided. The message was clear: in many instances, improvement made better sense than clearance although it was still recognized that in most cases this simply prolonged the life of property and postponed the day when clearance would be necessary.

The crucial change in attitudes came in 1967. Until then, government had relied for its assessment of the scale of the housing problem on estimates made by local authorities each using its own interpretation of the legal definition of unfitness. These seemed to show that the problem was fairly sharply concentrated in a few of the larger cities: a quarter of all unfit houses were thought to be in just four cities (Liverpool, Birmingham, Manchester, and Sheffield). That same report recognized that unfit dwellings in those cities would take forty-one years to clear at current rates of demolition and that, therefore, different grades of improvement would be essential to provide tolerable conditions to residents of these unfit dwellings in the interim (U.K. Ministry of Housing and Local Government, 1966b).

Within a year came the results of the first National Housing Condition Survey. This showed that the scale of unfitness was greater and more widespread than had previously been realized (2.9 million dwellings compared with under one million in the previous estimates). Even that number was greatly exceeded by the number of supposedly fit dwellings lacking at least one basic amenity or requiring significant repairs (4.5 million). In short, the survey suggested that one-sixth of the dwellings in England and Wales were unfit and a further quarter were in need of attention.

These facts were sufficiently serious for government finally to give its fullest support to a policy of improvement. Slum clearance programs were clearly unable to cope with the problem and were unable by themselves to arrest the progressive deterioration of fit dwellings into slums. Furthermore, the government White Paper stated that emphasis in the new policy should not simply be on the improvement of individual houses, but local authorities should "direct [their] main efforts . . . to the improvement of whole areas" (U.K. Ministry of Housing and Local Government, 1968).

The case for the improvement of houses by this time was well proven. But the emphasis on area improvement in housing policy was both new and, to a large extent, untested. The 1964 Housing Act had given local authorities the power to "declare" (i.e., designate) so-called *improvement areas* where more than 50% of dwellings lacked basic amenities but could be brought up to an acceptable condition by use of standard grants. The town and country planning acts had for a long time given powers to local authorities to declare *comprehensive development areas,* though these had been used chiefly for the redevelopment of town centers. The small area dimension had been given a strong impetus in the recommendation for so-called *action areas* in the highly influential report of the Planning Advisory Group (U.K. Ministry of Housing and Local Government, 1965), but there was very little actual experience with the comprehensive improvement of housing areas including dwellings and their environment. One of the few attempts, at Rye Hill in Newcastle, had proven to be a very mixed blessing in that neither the legislative powers nor the techniques had been readily available or understood (Davies, 1972). Nevertheless, the 1968 White Paper called for a change in emphasis in housing policy to make area improvement a key tool.

GENERAL IMPROVEMENT AREAS AND THE 1969 ACT

Area improvement thus became a part of the government's housing policy in the 1969 Housing Act. The new policy marked a decisive break with the slum clearance policy although the momentum of the clearance program was such that the actual number of dwellings cleared each year did not begin to fall significantly until 1974. The five years, 1969-1974, thus were a period of changing directions but nevertheless one in which policy was still equivocal on the crucial question: was improvement a real alternative to clearance, or a complementary program for maintaining the standard of older houses pending their eventual clearance?

The basic principle of the housing improvement grant system under the 1969 act continued to be voluntary action. Owners of houses (landlords or owner-occupiers) continued to be able to apply for a grant equal to 50% of the approved, eligible cost of improving, modernizing, converting or repairing their dwellings (U.K. Ministry of Housing and Local Government, 1969a). The grants were renamed but followed the earlier pattern: (1) *Improvement grants,* at the discretion of the local authority, were payable for a high level of improvement which would ensure a life of at least thirty years for the dwelling. For the first time, however, up to half of the grant could be for repairs and the replacement of outworn fittings. The conversion of larger houses into modernized, self-contained flats was encouraged by setting a higher ceiling for assessing eligible costs for grant assistance. (2) *Intermediate grants,* the old standard grant, were payable for installing basic amenities in houses with a life of at least fifteen years but less than thirty. (3) *Special grants* were payable for the improvement of houses in multiple occupation. Central government would meet 75% of the local authorities' expenditure on all three grants.

The system still relied on owners to come forward and apply for a grant. The local authority confined its activities to promoting the policy, stimulating the takeup of improvement grants and administering the complex procedures for grant approval. They had, but seldom used, reserve powers to compel unwilling owners (usually landlords) to repair or improve their houses.

The area dimension came in the guise of a new concept, the *general improvement area* (GIA). The explanatory advice from the central government (U.K. Ministry of Housing and Local Government, 1969b) asserted that a better return would be obtained if effort and resources were directed to extending by many years the remaining life of whole areas rather than individual houses. The characteristics of such areas were (1) an absence of any early proposals for major developments; (2) a clear potential for improvement, the buildings being structurally sound but lacking amenities or in poor repair and the environment capable of upgrading; (3) a favorable attitude towards self-improvement on the part of residents and owners. The areas should not be "too good," that is, already undergoing spontaneous improvement; nor should they be "too bad" in the sense that clearance was imminent. In the latter type of area, intermediate grants would be available on an individual basis where social needs required. But, usually, houses in GIAs would be eligible for full improvement grants. It was thought that GIAs would contain between 300 and 500 dwellings to ensure a reasonable degree of impact and to be administratively practicable. They should have well-defined physical or environmental boundaries.

A major effort would be made to build confidence in the future of each GIA. One method would be through extensive public participation, including exhibitions, public meetings, house-to-house visits, demonstration houses, and newsletters. Residents would be encouraged to form local groups (if they did not already exist) which would participate in the planning and management of improvement programs.

In addition, local authorities would demonstrate their visible commitment to the GIA by a program of environmental improvements which typically would include traffic management, car parking, and pedestrianization; providing play spaces, tree planting, and landscaping; and generally cleaning up the area and putting it in good order. Local authorities were permitted to spend up to an average of £100 per dwelling on this work, including the acquisition of land, and on this they would receive from central government assistance equal to 50% of the expenditure.

The commitment and the participation were intended to stimulate the greatest possible takeup of improvement grants, but neither the ceiling on eligible expenditure nor the rate of grant would be higher in GIAs than elsewhere. The only stimulus would come through the environmental works and the assurance of long-term freedom from clearance and redevelopment.

Administering a GIA would involve many different political committees and executive departments of the local authority (legal, financial, housing, environmental care), of other local authorities (county planning and highways), and of other agencies (public utilities). It was recognized that both the complexity of administration and the requirement for public participation would present new challenges to the traditional, highly compartmentalized, and paternalistic methods of local government, but little advice was given on how these might be overcome.

Indeed, a distinctive feature of the whole policy was the generally permissive stance of the Ministry of Housing and Local Government and later the Department of the Environment.[4] This was in sharp contrast to the much more restrictive and interventionist stance by government in other areas of housing policy, such as in the construction of council housing or control of rents in the private sector. It has been argued that the permissive approach reflected the relatively apolitical nature of the housing improvement policy at that time (Roberts, 1976).

As a result, government placed remarkably few constraints on local authorities. They had considerable discretion to interpret the rules about eligibility of costs for grant assistance. Ground rules for the selection of GIAs were few in number, none of them was mandatory, and in any case the government had no legislative power either to approve or reject the local authorities' choice of areas. No limits were placed on total expenditure on improvement grants and environmental improvements, provided individual grants lay within the ceiling figures. Instead, government issued advice ranging from the prototype case study of a GIA, at Deeplish in Rochdale (U.K. Ministry of Housing and Local Government, 1966a) to the series of Area Improvement Notes covering survey methods, housing and environmental improvements, the design of streets and open spaces, public participation, traffic, and management networks (U.K. Department of the Environment, 1971-1978).

About 1.3 million improvement grants of all kinds were made between passage of the 1969 act and the end of 1974 (U.K. Department of the Environment, 1977), but very few of these were attributable to the 911 GIAs declared during this same period. Within these, only 75,000 grants were made, or about 6% of the total (U.K. Department of the Environment, 1978).

The total annual number of grants increased considerably during the period 1971-1974, following a change in the ceiling figures for eligible expenditure but, more significantly, in the rate of grant itself. For two years (later extended to three) the rate was increased from 50% to 75% in the northern, northwestern, and Yorkshire-Humberside regions of England and in Wales. At the same time the levels of central government assistance to local authorities in these regions were increased, to 90% of expenditure on improvement grants and 75% of expenditure on environmental improvements in GIAs. The reasons for increasing the grants lay more in regional economic planning than housing policy, for the assisted regions were those experiencing relatively high unemployment. Still, the combination of increased grants and exchequer assistance for a known, brief period proved decisive.

The number of GIAs declared between 1969 and 1974 and the takeup of improvement grants within GIAs were both probably less than expected although no forecasts had been made. Two surveys of the program have been made, one in an academic study (Roberts, 1976) and the other by a government department (U.K. Department of the Environment, 1978). Both relied on surveys of a sample of GIAs and they agree fairly well in their conclusions, although the department survey includes post-1974 experience. The next few paragraphs draw on these two sources.

The average size of a GIA (312 dwellings) was just within the recommended range, but the actual spread of sizes was very great (between nine and 2,800 dwellings), and they covered a much wider variety of conditions than expected. On average, two-thirds of the dwellings in GIAs needed improvement or repair. The department survey noted that 43% of the dwellings in its sample needing improvement had actually been improved by 1976, of which 42% had been privately improved with grant aid, 40% were council dwellings improved with grant aid, and 18% had been privately improved without grant aid.

The high figure of council houses improved points to a major unanticipated use of the GIA policy. One-fifth of all GIAs were council housing estates from the interwar period, none of which had been expected to fall within the program. The houses may have been old-fashioned with poorer amenities than postwar housing. They (and their environment) may have been neglected and poorly maintained. But in no sense were these council estates typical of the older housing intended for the improvement program. However, central government had no power to stop declaration of GIAs, nor to stop local authorities using the exchequer assistance under the 1969 act and the even greater assistance following 1971. The reasons for their selection were fairly clear, even apart from questions of need and the availability of finance. Being in a single, local authority ownership, very often with a very restricted range of house types, council estates could be improved comparatively easily. Indeed, 86% of local authority GIAs had substantially completed their housing and environmental improvements by 1976 compared with only 11% of the private sector GIAs.

The private sector GIAs were very much more varied, chiefly because of the unwillingness of the local authorities to follow the strict letter of advice from the Department of the Environment and the inability of the department to enforce its advice. Nevertheless, the most successful GIAs were probably those which most closely corresponded with the original conception of the 1969 Act. One of the most fully documented cases was in Liverpool. Shelter, the national organization concerned with problems of poor housing and homelessness, set up the Shelter Neighborhood Action Project (SNAP) in a section of inner Liverpool. Between 1969 and 1972 SNAP formed residents' groups, prepared a fully documented survey report, set up a housing advice center and laid all the groundwork for a formal GIA. In the event, the local authority refused to act on the SNAP report and instead undertook its own surveys. Despite this false start, the area was soon declared a GIA and considerable progress has been made in housing and environmental improvements and in stabilizing the social and physical fabric of an area previously in decline (Shelter Neighborhood Action Project, 1972).

A more usual pattern is illustrated by the case of de Beauvoir Town, in Hackney, London, significantly an area thought to be of such architectural and historical interest as to be a conservation area under the 1967 Civic Amenities Act. Very little money is available under that legislation for the care and maintenance of buildings, the main powers lying in stricter planning control to conserve the area's character. But the local authority, in common with many others, used the grant aiding and environmental powers of the 1969 Housing Act to reinforce the planning powers available in a conservation area. And in this case the area was declared a GIA after a very fruitful public participation process (U.K. Department of the Environment, 1973a).

Environmental improvement presented problems. The initial allowance of £100 per dwelling proved to be too small, but in 1972 it was raised to

£200, a figure more in accord with the actual, final level of expenditure (U.K. Department of the Environment, 1976b). The major problems lay elsewhere. One lay in the previous neglect of the environment of older housing (U.K. Department of the Environment, 1977a). Another was the time taken to get the work in hand. The preliminary stages took from six months to five years between declaration of the GIA and the start of environmental works. The delays were caused by the complexity of the statutory procedures for street closures, traffic management, pedestrianization schemes, or the acquisition of land, say for play spaces or off-street parking. They were made worse in the early years by the preference of many local authorities for "once and for all" schemes of comprehensive environmental improvement (U.K. Department of the Environment, 1978).

Nevertheless, a departmental report could state that "it is the attractiveness of the house fronts and gardens which offers the clearest sign of a return of confidence to the area and it is the quality of the environmental improvements and their subsequent upkeep that provides the evidence that the area approach to improvement has worked" (U.K. Department of the Environment, 1976b). It goes on to say that the key to success is to "remove what offends, enhance the inherent qualities, and provide the lacking amenities." Despite these overt signs, visible in many GIAs, there was no firm proof that environmental improvements as such resulted in an increased takeup of improvement grants. Only one local authority (Cambridge) claimed to have found evidence that following declaration and the completion of environmental improvements, the rate of takeup of improvement grants had increased (U.K. Department of the Environment, 1978).

The geographical distribution of GIAs more or less followed the pattern of housing problems revealed in the 1967 National House Condition Survey: a third were in the conurbations, a half in smaller towns and a fifth in rural areas. Most of the more successful GIAs were in the last two types of area. The problem cases were located mainly in the conurbations where they were either "too good" or "too bad" to be appropriate for selection as GIAs. The former included areas already undergoing spontaneous improvement; declaration as a GIA merely accentuated the process. The process became known as "gentrification": the original residents, usually tenants, were displaced by more wealthy people who either bought a small house and improved it, or bought a converted flat in a subdivided, modernized large house. Gentrification was mainly a London phenomenon. A 1972 survey by the Department of the Environment showed that 75% of their original occupants moved when houses were improved with grant assistance, and that 80% of the movers were tenants (U.K. Department of the Environment, 1977c).

GIA status and even the improvement grants themselves arguably were neither essential nor equitable in areas being improved by gentrification. A study of Islington showed that its location, just north of the commercial center of London, was the crucial factor in stimulating its revitalization, allied with its Georgian character, then becoming fashionable, and changes in the rent acts. The pattern of tenure changed to owner occupation, financed initially by local building societies and later by the larger, national building societies relaxing their previous policy of refusing applications for mortgages in the area. Improvement grants and, more particularly, GIA status were almost irrelevant (Williams, 1976).

In other cities, GIAs and improvement grants could be crucial and, together with availability of private mortgage finance, could ensure the success of an improvement policy in the favored sections of a city such as Newcastle where, fundamentally, demand for older housing was low (Boddy, 1976). Elsewhere in the city, building societies were less willing to lend so the success of a GIA depended on the willingness and ability of the local authority to provide loans and mortgages. However, these were relatively plentiful up to 1974.

The more serious problem was in the housing stress areas, where high rents were being paid for poor-quality, privately rented housing usually in multiple occupation by poor tenants. This was chiefly a London phenomenon though also found in some inner areas of the larger conurbations such as Liverpool (Henney, 1973). The voluntary principle would not work in stress areas because of the reluctance of landlords to improve their properties, but the use of compulsory powers was cumbersome and uncertain. Strictly, such areas were too bad for effective GIA action, but in many cases it was politically impossible for them to be excluded from the program.

By 1973, the internal contradictions of the area improvement policy were becoming all too clear. An area selective policy with limited powers was being used to tackle a widespread problem, relying on a voluntary principle not even reinforced by higher rates of grant. The policy was successful in areas where the action corresponded with the intentions of the legislation and departmental advice. But in too many areas it was either misused for the improvement of interwar council housing estates; or was inequitable in its use of resources, where it reinforced a self-generating process of improvement; or, most seriously, was ineffective, in failing to tackle the problems of improvement in areas of housing stress.

Arguments were building up for local improvement policies which would cover the entire older housing stock of a city with a range of appropriate, area-based programs. The 1969 act had indeed laid a duty on all local authorities to survey their districts and define programs for clearance and improvement, but a wider range of powers was needed to resolve the contradictions of the current policy. This happened when government accepted the notion that improvement could be an alternative, and not simply complementary, to clearance; that other forms of selective area improvement were needed to reinforce GIAs which themselves should be more sharply focused; and that changing attitudes to housing tenure would be necessary if such policies were to be effective in areas of housing stress (U.K. Department of the Environment, 1973b).

HOUSING ACTION AREAS AND THE 1974 ACT

The key words of the 1974 Housing Act and its associated ministerial advice were "gradual renewal." The aim was to transform run-down residential areas into "decent, civilized neighborhoods" through a comprehensive strategy for the older housing, as a whole, by concentrating resources and effort into the areas of real need where additional powers would be made available (U.K. Department of the Environment, 1974a). Clearance and redevelopment were retained from previous legislation though they would be used more sparingly as the main 1960s clearance program came to an end. GIAs were also kept more or less intact from previous legislation, except that the Secretary of State for the Environment now had the power to prevent the declaration of individual GIAs if they failed to accord with policy guidelines from the central government.

Two new types of housing area were defined in the legislation: *housing action areas* (HAAs) and *priority neighborhoods* (PNs). Both would be declared by local authorities in much the same way as GIAs but there the similarities ended (U.K. Department of the Environment, 1975b). HAAs were intended to grapple with the problem of housing stress caused by the combination and interaction of bad physical and social conditions to which GIAs had been unable to respond. They were expected to be smaller than GIAs (about 200-300 dwellings) and the tenure at declaration would mainly be private rental. Council housing estates were explicitly excluded. The really distinctive feature of HAAs, however, was that each was to be purely an interim measure providing for action to be completed within five years to improve housing conditions, the well-being of the existing residents and the effective management and use of the housing stock. Once the immediate causes of the housing stress had been eradicated and the situation stabilized, an HAA could well be redefined as a GIA for long-term improvement. The physical action within an HAA, as in a GIA, would still be through grant-aided improvement of houses and environmental works for which, however, a smaller allowance was made than in the longer term GIAs (£50 per dwelling compared with £200).

The teeth in the HAA legislation came in the powers for enforcement and in the intentions for tenure. Local authorities were given much stronger powers to acquire properties if their owners or landlords, especially those with dwellings in multiple occupancy, were unable or unwilling to improve or repair them, or left them vacant. Authorities were empowered to collect information on ownership, and to be notified of any impending house sales. The intention was clear: privately rented property which was not properly managed, improved, and repaired would be taken into public ownership. One course would be municipal ownership but the more usual would be through housing associations, the nonprofit-making voluntary organizations financed by the Housing Corporation, which were to play a key role in relieving stress in HAAs and GIAs. Accordingly, the 1974 act gave additional power and support to housing associations registered with the Housing Corporation (U.K. Department of the Environment, 1974b).

HAAs thus reinforced the area improvement policy initiated by GIAs. PNs were introduced as an intermediate stage between the two to prevent further deterioration and a rippling effect from stress areas declared as HAAs. But the powers made available were less than in either HAAs or GIAs and in practice very few have been declared.

The area dimension was further reinforced by changes in improvement grants. Previously, improvement, intermediate, and special grants had been made at the same rate in all types of housing areas, and very few constraints had been placed on the offer of grant. The 1974 act changed this. Grants were to be made at different rates: 75% in HAAs with a possibility of 90% grants in individual cases of hardship; 60% in GIAs; and 50% elsewhere. At the same time, the offer of a grant was to be subject to more stringent conditions. It could be made only for houses below a specified rateable (property) value. And the grant would have to be repaid under certain conditions if the recipient sold the house (if an owner-occupier) or did not offer it for rental (if a landlord) within a period of five years (U.K. Department of the Environment, 1974a). A new repairs grant was added to the battery of assistance available in HAAs and GIAs.

Perhaps the most important point about the whole system was that authorities were now to be more flexible in awarding grants. For instance, in the special hardship provision in HAAs, grants could be paid, as it were, in stages with repairs and basic amenities at first and full improvement at a later date either by a new occupant of the house or after environmental works had been completed. Indeed, substandard but fit dwellings could remain unimproved if their occupants wished.

GIAs now were seen explicitly as an alternative to clearance, part of the package for gradual renewal which rejected not only the idea of comprehensive redevelopment but also any thought of crash programs. HAAs (and, to a lesser extent, PNs) were there as an intermediate stage, halting the decline of otherwise doomed neighborhoods and establishing a threshold for long-term action through GIAs. Clearance would be confined to very small groups of houses. The essence lay in gradual renewal as a "continuous process of minor rebuilding and renovation which sustains and reinforces the vitality of a neighborhood . . . responsive to social and physical needs" (U.K. Department of the Environment, 1975a).

The basis was thus laid in the 1974 act for an effective comprehensive policy for the improvement of older housing. A further 306 GIAs and 272 HAAs were declared between 1974 and 1978. The GIAs conformed broadly to the new requirements: they no longer included council housing estates or housing stress areas, or areas with a high rateable value. Otherwise, they were broadly similar to the earlier GIAs (U.K. Department of the Environment, 1978). The HAAs also conformed to the legislative intentions in that they were predominantly in areas of housing stress: 69% of all HAAs contained at least some multiple occupation; 53% of all dwellings were privately rented, 84% in disrepair and 44% unfit, though not necessarily to be cleared (U.K. Department of the Environment, 1976b).

Most significantly, two-thirds of all HAAs were in areas which had been blighted by the effects of previous housing legislation, having been either clearance areas within which demolition had been deferred or was incomplete, or in areas of extreme uncertainty, chiefly through their categorization as "short life" areas. A major contrast existed between the London and the provincial HAAs. The former were much larger (more than 500 dwellings) with all the classic symptoms of housing stress, indicated by the high incidence of private rentals and multiple occupation. The latter were smaller (fewer than 200 dwellings); many were former clearance areas and progress in declaration was more rapid than in London.

The new rules resulted, too, in changing the distribution of improvement grants from privately rented (and, to a lesser extent owner-occupied and council) dwellings in favor of housing associations. Before the 1974 act, only 2% of grants were being made to housing associations, but by 1977 the proportion was 16%. However, grants within HAAs and GIAs showed no real change, staying at a combined figure of about 7% of the total annual number of grants between 1974 and 1976.

The total number of grants made each year declined rapidly, from over 230,000 in 1974 to just half that number in 1975, from which time it has remained constant (U.K. Department of the Environment, 1979a). In part, this decline is a result of the cuts in public expenditure since 1974 which have affected, in turn, the availability of local authority mortgages and loans for improvement essential for promoting owner occupation in older housing, and the finance for improvement grants, advances to housing associations, and local authority staffing. During the economic recession, the relative prices of unimproved and new houses and the costs of improvement also changed. Building costs, in particular, rose rapidly while the percentage of total improvement costs covered by grant aid tended to fall (Lomas and Howes, 1977). As a result of this, and the shifts in house prices, a so-called "valuation gap" opened up such that the price of an unimproved house and the cost of its improvement to the individual was greater than the market price of the improved house. This created difficulties in raising mortgage finance and reduced the incentive for improvement (Harrison, 1977; U.K. Department of the Environment, 1978).

The general economic reasons for the decline in the number of improvement grants were reinforced by specific changes within the policy itself. The ceiling on rateable value was intended to cut the number of grants made for the improvement of more expensive houses. The conditions on resale or letting were intended to prevent speculative gains, although evidence from Birmingham, for instance, showed that the conditions also ruled out a significant proportion of improvable dwellings because they were held on short leases (Freeman, 1977).

A third change in policy was crucial in reducing the number of grants. The policy was now intended to favor HAAs and GIAs by the offer of higher rates of grant within them. But it left virtually untouched the procedures for declaring areas, administering programs for environmental improvement, public participation, and grant approval. And it added a very substantial amount of staff work for planning, enforcement, and compulsory acquisition of dwellings in HAAs. The Birmingham study showed that for an average size area (say 250 dwellings), an HAA would require about 270 man-months of staff work and many more committee meetings, compared with 20 man-months for a GIA (Freeman, 1977). Yet the requirement for extra staff came at a time when public expenditure cuts were building pressure for reductions in staff levels (Howick, 1977).

The net effect was clear. HAA programs were already falling into arrears, which effectively ruled out the achievement of their objective or arresting decay by quick action. Estimates for Bristol, for instance, identified twenty-five–twenty-eight potential HAAS which, at current rates of progress, would take sixteen years to complete (Short and Bassett, 1978). Political questions about priorities and ways of structuring the program became important under these circumstances. Some cities, such as Birmingham, made known their intention to undertake a very large program of twenty-three HAAs and seventy GIAs in order to build an irreversible political commitment, but they encountered considerable opposition (Paris, 1977). Other cities decided to adopt more modest statutory programs which were within their capability, and used various nonstatutory devices such as "improvement zones" (Leeds) or "long-term revitalization areas" (South Tyneside) within which security from clearance was offered, to encourage the takeup of improvement grants elsewhere (Gibson and Longstaff, 1977).

AREA IMPROVEMENT AND THE INNER CITY

The condition of the older housing stock visibly improved between the first and third National House Condition Surveys in 1967 and 1976. The number of unfit dwellings was reduced by 50% and the number of dwellings lacking one or more of the basic amenities by 59%. Only disrepair showed an increase of perhaps one-third, allowing for inflation (U.K. Department of the Environment, 1977c). This achievement was gained by an excess of slum clearance (0.5 million dwellings) and grant-aided improvement (1.7 million) over the deterioration of dwellings into unfitness.

Area improvement policies played a comparatively small part in this process. By the third quarter of 1978, 1,218 GIAs and 272 HAAs had been declared, containing about 440,000 dwellings, not all of which needed improvement or repair. Surveys showed that even in the best year (1975) only 10% of improvement grants were made in the two types of area (U.K. Department of the Environment, 1979a). But any assessment of the area improvement policy is bedeviled by a confusion of goals between improving the national housing stock and the more local objective of revitalizing older neighborhoods for the benefit of their existing residents (Roberts, 1976).

It is not difficult to show the apparent irrelevance of the area improvement policy in improving the national housing stock. The declared HAAs conform reasonably well to the intentions of the act but various estimates have shown that more than 2,000 would be required to do the job on the basis of 1971 census indicators (Short and Bassett, 1978; Lawless, 1977). The estimates are open to question because of the length of time since the census, when improvements have been at a peak rate, and on the validity of using census indicators (Dennis, 1978). Nevertheless, the general point is well made. Unless it can be shown that GIAs and HAAs lead to a dramatic increase in the rate of takeup of improvement grants, they are not a cost-effective tool for improving the national housing stock. And unless many more HAAs can be declared quickly, the choice of a few raises questions of equity of treatment for the many similar dwellings excluded from the program.

The success of the social objectives must be judged in the light of improved conditions within each GIA and HAA. The department's own study argues that area improvement policies can be effective when operated in the spirit and intentions of the legislation. Individual neighborhoods have been made better places to live, with improved relationships between the local authority and residents and an enhanced community spirit. The program has been "accepted by local authorities as a significant and useful means of improving the quality of the nation's housing stock" (U.K. Department of the Environment, 1978).

Area improvement thus needs to be judged by local criteria as well as by its potential contribution to national policies. The risks lie in the reliance on essentially short-term objectives, on the principle of voluntarism, and the danger that, as an area selective approach, it misses too many of those in greatest housing need (Derrick, 1976). The advantages are its flexibility, giving it the capacity to respond sensitively to local conditions. The tragedy is that, just as area improvement is gaining a clear rationale as part of the concept of gradual renewal of the older residential areas, the whole policy has been thrown off course by the economic recession. It is too early to say whether the lost ground can be recovered and the policy of gradual renewal made to work, or whether conditions will deteriorate to a point where the only solution will be a resumption of slum clearance on a massive scale.

NOTES

1. The terms Britain and British are used for convenience but this account deals with policies for older housing only in England and Wales. The housing stock, its pattern of tenure, and housing legislation are different in Scotland from the rest of Great Britain. All statistics and references are to England and Wales unless otherwise stated.

2. This chapter draws heavily on the results of a recent major review of housing policy carried out by the government. The results of the review were presented in a consultative document which briefly covered the ground and identified the major

issues for housing policy in the future (U.K. Department of the Environment, 1977b). It was accompanied by three volumes of technical survey and analysis of the housing stock, its condition and tenure (U.K. Department of the Environment, 1977c). Chapters 1 and 2 in Part I, on the historical background and the current housing situation, and chapter 10 in Part III, on obsolescence and improvement in the housing stock, are of particular relevance.

3. Standards for new housing construction are defined in precise technical terms in legislation and regulations. Standards for the enforcement of slum clearance and housing improvement are much less precise. *Unfitness* is defined for the purposes of slum clearance in section 4 of the 1957 Housing Act. It is a list of factors to be taken into account but for which precise measures are not defined. The factors include stability, ventilation, freedom from damp, and provision of amenities. Other aspects of housing policy, including housing improvement, rely on other criteria including *lack of basic amenities* (fixed bath, indoor toilet, wash basin, sink, hot and cold water supply); *overcrowding* (more than 1.5 persons per habitable room); and *poor repair.*

4. The Ministry of Housing and Local Government was responsible for housing and planning in the central government until 1970, when it, the Ministry of Transport, and the Ministry of Works were amalgamated to form a new Department of the Environment. The structure was changed once more in 1976, when the Department of Transport was recreated as a separate ministry.

REFERENCES

BODDY, M. J. (1976) "The structure of mortgage finance, building societies and the British social formation." Transactions, Institute of British Geographers (New Series) 1, 1: 58-72.

BURROWS, J. (1978) "Vacant urban land." The Planner 64 (January): 7-9.

DAVIES, J. G. (1972) The Evangelistic Bureaucrat. London: Tavistock.

DENNIS, N. (1972) Public Participation and Planners' Blight. London: Faber & Faber.

――― (1978) "Housing policy areas: criteria and indicators in principle and practice." Transactions, Institute of British Geographers (New Series) 1, 1: 2-22.

DERRICK, E. F. (1976) "House and area improvement in Britain: bibliography and extracts, summary report and recommendations." Research Memorandum 54. Birmingham: Centre for Urban and Regional Studies.

FLEETWOOD, M. (1978) "Selling the inner city: the case of Stanfield Road." Architects' Journal 168 (July 12): 72-79.

FREEMAN, J. (1977) "Small Heath, lessons from an inner area study." The Planner 63 (March): 46-47.

GIBBONS, A. A. (1977) "Valuation and the inner city areas." Presented at a conference of the Royal Institute of Chartered Surveyors (Planning and Development Division), London, May 20.

GIBSON, M. and M. LONGSTAFF (1977) "Scope for improvement: policies and strategies in gradual renewal." The Planner 63 (March): 35-38.

HARRISON, A. (1977) "The valuation gap: a danger signal?" Centre for Environmental Studies Review 2 (December): 101-103.

HENNEY, A. (1973) "Managing older housing areas." Journal of the Royal Town Planning Institute 59 (February): 73-77.

HOWICK, C. (1977) "Manpower in local authorities: the joint manpower watch." Centre for Environmental Studies Review 1 (July): 11-18.

LAWLESS, P. (1977) "Housing Action Areas: powerful attack or financial fiasco." The Planner 63 (March): 39-42.

Liverpool Shelter Neighborhood Action Project (1972) Another Chance for Cities SNAP 69/72. London: Shelter.

LOMAS, G. and E. HOWES (1977) "Private sector improvement since the 1974 Acts: some facts and figures." Centre for Environmental Studies Review 2 (December): 109-112.

MASON, T. (1977) "Inner city housing and renewal policy: a housing profile of Cheetham Hill, Manchester and Salford." Research Series No. 23. London: Centre for Environmental Studies.

NEEDLEMAN, L. (1965) The Economics of Housing. London: Staples Press.

PARIS, C. (1977) "Birmingham: a study in urban renewal." Centre for Environmental Studies Review 1 (July): 54-61.

ROBERTS, J. T. (1976) General Improvement Areas. Farnborough: Savon House/ Lexington Books.

SHORT, J. R. and K. BASSETT (1978) "Housing action areas: an evaluation." Area, Institute of British Geographers 10, 2: 153-157.

U.K. Department of the Environment (1971) Teesside Survey and Plan: Final Report, Volume II: Analysis, Part I, by Hugh Wilson and Lewis Womersley and Scott Wilson Kirkpatrick and Partners (Consultants). London: H.M.S.O.

––– (1971-1978) Area Improvement Notes 1-11. London: H.M.S.O.

––– (1973a) "Public participation in general improvement areas." Area Improvement Note 8. London: H.M.S.O.

––– (1973b) "Better houses: the next priorities." Cmnd 5339. London: H.M.S.O.

––– (1974a) "Housing Act 1974: improvement of older housing." Circular 160/74. London: H.M.S.O.

––– (1974b) "Housing Act 1974: housing corporation and housing associations." Circular 170/74. London: H.M.S.O.

––– (1975a) "Housing Act 1974: renewal strategies." Circular 13/75. London: H.M.S.O.

––– (1975b) "Housing Act 1974: housing action areas, priority neighborhoods and general improvement areas." Circular 14/75. London: H.M.S.O.

––– (1976a) "Environmental improvements: a report on a study of the problems and progress of environmental improvements in a sample of general improvement areas," by A. Williamson. Improvement Research Note 1-76. London: Author.

––– (1976b) "Housing action areas: a detailed examination of declaration reports." Improvement Research Note 2-76. London: Author.

––– (1977a) "Inner area study of Liverpool: environmental care project," by Hugh Wilson and Lewis Womersley (Consultants). London: Author. (1 AS/L1/19).

––– (1977b) "Housing policy: a consultative document." Cmnd 6851. London: H.M.S.O.

––– (1977c) Housing Policy: Technical Volumes, Parts I, II, and III. London: H.M.S.O.

––– (1978) "General Improvement Areas 1969-1976: a report on a comprehensive study of progress achieved in the rehabilitation of older housing in general improvement areas," by A. Williamson and E. Wrigley. Improvement Research Note 3-77. London: Author.

—— (1979) Housing and Construction Statistics No. 27, 3rd quarter 1978. London: H.M.S.O.

U.K. Ministry of Housing and Local Government (1965) The Future of Development Plans: A Report by the Planning Advisory Group. London: H.M.S.O.

—— (1966a) The Deeplish Study: Improvement Possibilities in a District of Rochdale. London: H.M.S.O.

—— (1966b) Our Older Homes, a Call for Action: The Report of the Subcommittee on Standards of Housing Fitness. London: H.M.S.O.

—— (1968) "Old houses into new homes." Cmnd 3602. London: H.M.S.O.

—— (1969a) "Housing Act 1969: house improvement and repair." Circular 64/69. London: H.M.S.O.

—— (1969b) "Housing Act 1969: area improvement. Circular 65/69. London: H.M.S.O.

U.K. National Economic Development Office (1971) New Homes in the Cities: The Role of the Private Developer in Urban Renewal in England and Wales. London: H.M.S.O.

WILLIAMS, P. R. (1976) "The role of the institutions in the inner London housing market: the case of Islington." Transactions, Institute of British Geographers (New Series) 1, 1: 72-82.

12

Urban Development:
An Alternative Strategy

EUGENE J. MEEHAN

☐ IN THE UNITED STATES, federal programs for improving urban conditions share a common origin in the massive extension of the accepted role of government in society that took place in the 1930s. The New Deal era placed an unmistakable and sometimes unfortunate imprint on American social programs that seems likely to remain strong through the twentieth century. In such fundamental areas as housing assistance, unionism, social security, unemployment security, and regulation of business practices, the principles, and even the legislation on which federal programs depend, have changed little since the late 1930s. Moreover, as the United States became increasingly involved with international affairs, the same assumptions and principles were applied to foreign economic assistance. The various development programs that have emerged out of the effort to achieve social purposes at home and abroad vary widely in substance and focus, but they share a number of common attributes that, taken together, make up an overall "approach" to development that is remarkably consistent.

The influence of that approach has been much strengthened by the inability of the political-administrative apparatus to improve performance and policy on the basis of experience. Few if any federal programs have operated as experiments, as occasions for learning and improvement. There is also very little evidence of sustained efforts to refine and improve performance over time, even in major urban programs. Indeed, programs seem at times to worsen rather than improve as the latent functions of the activity acquire greater significance than the ostensible purpose—public housing is a good case in point (Meehan 1975, 1979a). While there are some notable exceptions, a pattern of very poor performance seems to hold across a wide range of activities with depressing regularity. The system may be "working," as optimists declare, but the evidence suggests that its operating efficiency is extremely low and is not improving.

The strengths and weaknesses of established practices tend to be most clearly revealed when they are seriously challenged. With respect to social legislation and development efforts, one of the clearest challenges has come in the area of foreign assistance. The Inter-American Foundation (IAF) stands out as one of the few federal efforts to break with established tradition since World War II (Meehan, 1979b). Indeed, to the extent that it attaches primary importance to a deliberate effort to learn by experimenting with nontraditional procedures, the legislation is revolutionary, and the foundation's operations serve as a microcosm in which alternative approaches to development can be created and tested. IAF is a public corporation, established by Title IV of the Foreign Assistance Act of 1969. The mandate to the foundation was remarkable: it was to create and test new approaches to foreign assistance working outside the established foreign policy apparatus as much as possible. The opportunity provided by the legislation has not been exploited as fully and systematically as might be wished, and the evidence is still tentative and incomplete. But the alternative approach to fostering and supporting social and economic development in Latin America that is evolving out of IAF operations is worth the attention of all those seriously concerned to improve the quality of governmental performance in these areas. Indeed, the approach is significant as much for the way it has evolved as for its substance.

THE TRADITIONAL APPROACH TO DEVELOPMENT

A brief summary of the traditional or established American government approach to development, concentrated on faults or weaknesses, will provide a baseline for assessing the potential and achievements of the alternative approach to development emerging from IAF operations in Latin America.

POLITICAL IMPERATIVES

Some of the major influences on American federal development legislation flow simply from the political imperatives of the society. Two are particularly important: first, the basic soundness of the social, political, and economic institutions of the society is taken for granted; second, no program will come into being unless it can obtain legislative, and in some cases, public support. In combination, these two factors account for many of the basic weaknesses in development programs.

If the existing institutional structure is taken as given, the overall approach to development is necessarily remedial rather than preventive. Those in government have usually construed their task in terms of providing auxiliary support, often temporarily, and not restructuring foundations. The federal government has sought ways of dealing with exceptions and aberrations rather than ways of altering institutional arrangements so as to reduce exceptions and aberrations. For example, it is clearly more appropriate to deal with minor aberrations in a complex system by direct transfer payments than to tinker with fundamentals. If the system is working, wisdom suggests it be allowed to continue.

The need to gain legislative and popular support for program proposals has had an equally profound impact on development programs. In principle, three different types of appeals for support could be made; in practice, only one has proven workable and it has had some very undesirable side effects. The first, a humanitarian appeal, based on considerations of equity and justice, has often attracted widespread emotional and rhetorical agreement. But neither the legislators nor the general public have shown any real willingness to accept the significant costs of major programs for humanitarian reasons. A second possibility, appealing to collective self-interest, has also proven ineffective. The appeal is simple: the long-run health and vigor of society, which is vital for all its members, require elimination (at cost) of social weak spots. Unfortunately, the argument involves a composition fallacy. It is clear that in some senses everyone in society is in the same boat, and if the boat sinks, to take the extreme case, death by drowning is likely to be the common lot. But it is also clear that a very large boat can tolerate extreme variations in the conditions of life of its passengers without floundering. In fact, one school of thought asserts that without some privation below-decks discipline could not be maintained, and the quality of life for everyone on board would be lower.

For the most part, those seeking development assistance from the federal legislature have had to rely on a third appeal, directed at individual or specific self-interest. Though this appeal has been effective, the result too often has been programs that are misdirected or grossly inefficient. The reward to those who support development legislation out of self-interest must be meaningful and even tangible to be taken seriously. And, since legislation at the national level requires support from a range of interests, program purposes must be very broad and flexible. Most development programs, domestic or foreign, have ended by including something for everyone, much reducing the benefit to the ostensible beneficiary. Public housing, to take a concrete example, was justified as a means of reducing unemployment, a device for clearing slums, a boon for the construction industry, and a way of satisfying a serious human need. Its provisions therefore, included measures that would provide major rewards to the interests involved. Such practices dilute the central thrust of any assistance program.

Furthermore, there is an almost universal belief in the superior ability of the private sector to "get things done." In consequence, public programs are usually limited to provision of services and subsidies; productive enterprises are off-limits to government unless a private agent can be interposed who profits from the activity. Exceptions such as the Tennessee Valley Authority are rare. Such built-in bias has made it very difficult for development programs to generate the momentum needed to continue once financial support is withdrawn and thus reinforced the tendency for social programs to foster dependency among their clients. Curiously enough, the use of social assistance to finance development of enterprises founded on individual profit-making is uncommon. Those who support development assistance tend to regard profit-oriented operations as incompatible with their purposes much as the profit-oriented tend to assign humanitarian considerations a peripheral place in the entrepreneur's obligations. It seems likely that both extremes are untenable in the long run.

PROGRAM CHARACTERISTICS

Few development programs, whether at home or abroad, have been particularly successful. The sources of failure lie, I believe, in certain fundamental characteristics of the programs. Among the more important of them are: (1) a macroeconomic theoretical focus, (2) centralized planning and administration, (3) a "brick and mortar" orientation, (4) inadequate resource allocations, (5) lack of participation and cooperation from program beneficiaries, and (6) lack of adequate machinery for improving performance over time on the basis of experience. Collectively, these factors have had a profound and unfortunate impact on development efforts, domestic or foreign.

Macroeconomic Focus

Both domestic and overseas development are construed by those in government in macroeconomic terms on the Keynesian model. The effect, most readily visible in foreign assistance, has been disastrous. If problems tend to be identified in terms of productive inadequacies rather than distributional skews, solutions are sought in measures intended to expand gross national products or national income. Foreign assistance, for example, has relied primarily on large-scale capital transfers, creation of major infrastructures. Programs usually involve heavy infusions of advanced technology and large-scale use of capital equipment, backed by modern organizational and management techniques. In domestic affairs, the macroeconomic commitment suggests such tactics as creation and renewal of capital equipment, support for large-scale infrastructure development, and heavy subsidies for the private sector. The basic strategy in both areas is to concentrate resources at the top of the economic structure and wait for the effects to "trickle down" to the rest of the population. As has been pointed out with respect to foreign assistance, that strategy virtually guarantees failure:

> In many ways these policies reflect a continuing and fundamental misunderstanding of the problem of poverty and imply a belief that if the national income can somehow be raised it will *ipso facto* raise the lot of all those contained therein. The truth that groups of people are differentially (and some even adversely) affected by a rising national average income is hardly a revelation to those who care to recall their reading of Dickens [Thiesenhusen, 1978: 161].

The failure of the "trickle down" strategy in domestic programs is equally clear though less frequently remarked; it is particularly easy to document in the field of housing (Nourse and Phares, 1974; Leven, et al., 1976; Phares, 1977). In relation to development, *how* increases in national income are achieved and distributed may be as important for the total population as the actual amount of increase.

Centralization

The social programs inaugurated in the 1930s were large-scale national ventures, planned at the top, and delivered through successive layers of public agencies to the ultimate consumer. At the time, centralization of policy-making and administration was probably essential. State and local authorities were unable to cope with the aggregate disaster brought about by private sector failure. National programs offered a way around the parochial and conservative outlook prevalent in state and local politics. For the short run, central planning, financing, and administration were unavoidable.

Over the longer run, centralization has had most unfortunate consequences for society. The two basic faults in centralized planning and administration are simple but lethal: first, they literally rule out systematic improvement of policy on reasoned grounds; second, they foster extreme client dependency.

It is commonplace that large-scale, centralized operations require massive bureaucracies. When there are a great many administrative layers between the policy-making authority and the ultimate client, it is impossible to transmit information accurately and adequately from the point of contact with the client to the policy-making center. In the process of aggregating the data, a conceptual transformation is required; that creates a conceptual gap between action and consequence that can only be bridged by a level of theory nowhere available in social science.

The difficulties created by unavoidable conceptual transformations are compounded by the bureaucratic effects of growth. Large-scale administrative agencies tend to be opaque, hence uncontrollable. Within the bowels of such monsters as HUD or HEW, congressional mandates are lost or transformed in unfathomable and unpredictable ways. To take a concrete example from the 1965 Housing Act, a congressional directive to support renovation of existing housing for lease was transformed into support for the construction of new housing for lease, an activity specifically ruled out in congressional discussions. Policies remain radically incorrigible so long as the administrative agency exerts a substantial but unknown influence on program outcomes.

Client dependency, the second undesirable effect associated with centralized operations, seems equally unavoidable. For reasoned policy-making, data relating to the specific effects of action on the individual client must be integrated with aggregate data; the machinery for such integration is very difficult to establish and maintain. The public hearing and advisory panel, which are the usual substitutes for close interchange between client and administrator, are rarely effective. The common practice of appointing administrators with no local experience (in the mistaken belief that there are "principles of management" sufficient for the conduct of public affairs that can be learned apart from the substance of an operating program) reinforces the separation. The optimum condition in which agency and client coordinate their efforts to achieve agreed purposes with maximum efficiency requires decentralization. Knowledge of the detailed effects of program activities must be obtained locally; without it, there can be no reasoned planning. Planning is necessarily a *local* function requiring central integration.

The "Bricks and Mortar" Orientation

To a degree surprising in a democracy, which must at least pay lip service to the primacy of persons, development programs are targeted at the production of physical capital. The Community Development Block Grant Program in St. Louis, for example, has provided funds for such activities as street improvements, sanitary sewers, storm sewers, cleanup of streets and lots, and demolition of vacant and dilapidated structures; neighborhood facilities, especially community centers. While it might be argued that other programs are available for "human" development, the fact remains that "community" development tends to be conceived as development of *physical* capital, isolated from the populations that development is expected to serve.

The primary effects of this overcommitment to physical capitalization appear in program evaluation; if purposes are stated in physical terms, achievements will be measured on the same scales. Yet physical improvements are not an intrinsic good, something valued for its own sake. Physical capital is valued for its effects on human populations. At best, the practice of concentrating on the physical is liable to serious abuse, as in the evaluation of school quality in terms of teacher salaries, teacher-pupil ratios, or expenditures per student rather than student performance.

Inadequate Funding

Serious efforts to deal with serious problems begin, necessarily, with a survey of the nature and scope of the problem. That leads to an assessment of the amount and kind of resources, technology, and so on, required for improving the situation. To my knowledge, no federal development or assistance program has evolved from such a survey. Of course, the survey is unnecessary if (a) the government intends to deal with the problem whatever the cost or (b) the government has no intention of dealing with the problem. In the latter case, public relations and distribution by lottery of inadequate resources can be expected. Indeed, inadequate funding is one of the best available indicators of governmental unwillingness to accept the real costs of dealing with social problems. If that unwillingness were openly acknowledged, which is very unlikely given the political climate, the harm done would be modest. But when significant programs are so grossly underfunded that failure is unavoidable yet the presumption is allowed to remain that the program has been given a fair trial and found wanting, the result is condemnation without trial and unwarranted denigration of potentially useful programs.

Again, public housing provides a perfect illustration. The fiscal arrangements incorporated into the Housing Act of 1937, and repeated in the Housing Act of 1949, were *predictably* lethal. The federal government paid the capital costs of constructing the buildings; other expenses had to be met solely from rents—there were no operating subsidies. Tenant incomes were deliberately kept very low. In consequence, local housing authorities had either to charge the tenants excessive rents or find themselves unable to operate and maintain their property. The problem was magnified by the requirement that local authorities pay 10% of rent to local government in lieu of taxes, provide utilities for each resident, and pay part of their capital costs with any overages remaining after a small reserve fund was established. The crowning irony appeared in 1969 when the first Brooke Amendment limited tenant rent to 25% of gross income but provided no subsidy for housing authorities to make up for lost revenue. By the end of 1969, housing authorities in virtually all American cities were facing financial ruin. To take their condition as evidence that "public housing" had been tried and failed, as frequently happened in Congress and elsewhere, was a travesty.

Program Performance

How well has the traditional strategy performed? The exact dimensions of the performance are unknown, which says something very important about the way the strategy was applied. On the whole, ineffectiveness has been matched only by expense. Curiously enough, recognition of the failure of the traditional approach appeared earlier and more clearly in foreign assistance than in domestic programs.

So long as discussion is restricted to aggregate physical and monetary terms, the achievements of past programs appear impressive. Enormous sums have been spent; a great many physical entities (roads, buildings, or airports) have been constructed, repaired, or demolished; hosts of people have been provided with short-term employment; enormous quantities of goods and services have been purchased and distributed. Unfortunately, when the scale is very large, the costs of error can also be very high, indeed. Countless billions of dollars have been spent in a futile and misdirected effort to "save the cities," for example, with little thought to just "what" was being saved, what "being saved" meant in context, or why such "savings" justified the enormous costs. Similar aberrations can be found in every area of development.

If program evaluation is extended to take into account the effects of action on the various economic strata of society, whether at home or abroad, the weakness of the traditional strategy is patent. Where physical facilities such as roads or airports have been constructed, the skew of user benefits to the higher income brackets is to be expected and may be justified. But when higher-income persons are also the principal beneficiaries of low-income housing subsidies and other programs ostensibly targeted at the needy, that is cause for concern.

What of the needy? The lucky winners in any federal lottery gain significant benefits. In the 1940s, for example, residents of public housing were better housed at less cost than almost anyone else in the poverty sector. Indeed, one of the more frustrating aspects of social legislation is the solid evidence of significant achievement where programs were given half a chance to succeed. Public housing has provided, and still provides, decent housing at an affordable price in decent surroundings for a great many persons who could not afford such accommodations on the private market. Unfortunately, it also provides inadequate housing in terrible surroundings for many others. Few federal development programs have been able to perform adequately for any extended period on a significant scale.

Beyond the discernible effects of action lie the uncertain effects of avoidance and neglect. The basic exploitative apparatus in society remains virtually untouched, for example. The usurious lender and the overcharging merchant remain, and the labor unions, whose activities triple or quadruple the cost of decent housing, new or rehabilitated, are as much exploiters of the poor as the local usurer or storekeeper. The needy continue to be the first and most helpless victims of inflation or the dislocation brought about by technological change. In Latin America, for example, commercialization of agriculture and overuse of pesticides have had a catastrophic impact on traditional subsistence farmers. In the United States, programs that provide benefits greater than those available in the local labor market serve to transfer populations from self-sufficient employment to dependency. Most important of all, the failure to provide consistent support for social organization, combined with the tendency for federal programs to corrupt existing organizations to the search for federal hand-outs, has had corrosive effects on the poor. In modern society, organization is the key to self-sufficiency as well as the best means available for obtaining transfer payments. The kinds of organizations that are needed to bring the poor into active participation in the mainstream of society are not available and seem not to be evolving or growing. Over the long run, such organizations are probably society's greatest need and their absence a measure of society's principal failure.

AN ALTERNATIVE STRATEGY IN FOREIGN ASSISTANCE

What is most striking about the failure of the traditional development strategy in domestic affairs is how seldom the failure is noted. The track record in foreign assistance is no better, but there it has been sharply attacked. With respect to Latin America, the performance of the Alliance for Progress and the Agency for International Development (AID) was being subjected to harsh criticism by the 1960s from the United States and Latin America alike. The principal target of the critics was the paternalism clearly visible in AID operations. As early as 1966, that led to incorporation of Title IX into the Foreign Assistance Act, requiring active participation in program operations by beneficiaries and support for development of indigenous democratic institutions. Title IX proved relatively ineffective. In 1969 the Subcommittee on Inter-American Affairs of the House Foreign Relations Committee, after prolonged hearings on foreign assistance, issued a major report. While the report reaffirmed the basic goals of foreign assistance—transformation of Latin American society, economic growth, elimination of disease and illiteracy. achievement of better income distribution, extension of social benefits, and development of free democratic institutions—it concluded that American foreign assistance had failed to move Latin America closer to these goals and might in fact have been counterproductive.

What made the weakness of the overall development strategy somewhat easier to see in foreign assistance was the identification of a well-defined target population. the very poor and needy members of society. The "trickle down" strategy had produced few significant improvements in their living conditions. Moreover, the resources available to the poor were significantly reduced by hidden subsidies to both American and Latin American interests. Finally, American policy was increasing both the scale and the intensity of dependence among the poor and not fostering self-reliance or capacity for self-sufficiency—a result that was generally attributed to the practice of working out country plans for each nation and implementing them using American planners, American technology, and very often, American organizations.

THE INTER-AMERICAN FOUNDATION

The congressional response to the strategic inadequacies revealed in the 1969 hearings was not a new strategy but an instrument to develop and test new strategies in Latin America, the Inter-American Foundation. IAF was supplied with a target population (the poorest segment of the Latin

American population) and given the task of learning how to convey assistance to them as effectively as possible. A few overall guidelines were set, mainly to avoid repetition of mistakes previously identified in AID operations and to eliminate some of the more serious impediments to learning associated with the operation of the domestic bureaucracy. Beyond that, IAF was free to chart its own course.

The alternative strategy that IAF has evolved is a product of experience and not of speculation. Admittedly, the evidence is often fragmentary: the "start-up" time was prolonged by the unusual circumstances; the quality of project records varies; many projects have not yet generated adequate evidence for judgment. Where the strategy has been applied systematically, it seems to have worked quite well. Occasionally, results have been spectacular. But it remains a promising alternative, a strategy worth further exploration and not a finished product awaiting routine application. Further, the strategy should probably be considered a supplement to and not a substitute for existing practice. IAF has operated in areas of great poverty with little socioeconomic infrastructure, concentrating on the needs of the very poor. Modifications will probably be needed if the strategy is used in more developed societies. In general, the relation between IAF-style operations and the wider social-political-economic system remains to be worked out. So long as the scale of operations remains small, such relations can probably be ignored—though there are already some signs of strain. A major commitment of resources to the IAF strategy would certainly require a close examination of those relations and their program implications.

The legislative mandate to IAF contained three basic operating principles: (1) IAF would remain independent of the regular foreign policy machinery and of the established bureaucracy, (2) the program would be carried out through direct relations with indigenous private organizations, (3) the agency would not make country plans on the AID model but would respond to initiatives from Latin America.

Independence from the bureaucratic establishment was perhaps the central concern of the legislators responsible for the new agency. To achieve it, IAF was organized as a public corporation with a seven-member Board of Directors, four of whom were appointed from the private sector. To eliminate the need to return to Congress each year for more funds, at least in the initial stages of operations, $50 million was transferred from the AID appropriation and made available until spent. Field operations were to be separate from both department of state control and interference by the U.S. embassy in the host country. Relations with the host country are the responsibility of the grantee; IAF supplies information to the host country embassy in Washington. The congressional effort to establish operating independence for the Foundation did not succeed completely; the Office of Management and Budget (OMB), for example, quickly asserted control over budgetary matters. But on the whole, IAF has operated with a freedom from interference rare in Washington.

The internal organization and operating procedures of the foundation have also evolved with experience. A field representative is assigned to a country or cluster of smaller countries; the countries are grouped by regions, each headed by a director. The entire staff is based near Washington, DC; field representatives travel frequently to their assigned areas but do not reside there. Applications for assistance, usually made by indigenous organizations, are considered first by the field representatives, then by a small committee. The final decision on funding is made by the regional director on advice from the staff. Contracts are relatively flexible; both the purposes of the grant and the means of attaining them can be changed. Review and coordination are obtained by periodic meetings among the regional directors and the foundation's president. The Board of Directors maintains overall supervision of operations.

LOCALISM OR BUILDING FROM BOTTOM UP

The approach to development that has evolved out of IAF experience is built around three basic elements: a human focus, recognition of the centrality of social organization, and responsiveness to external initiative. In combination, they imply a strategy radically different from that followed by AID or in most domestic or foreign assistance programs.

Identification of a definite target population. The poor and very poor in Latin America were the target population for IAF efforts. And since IAF had a limited budget, and efforts to deal with the poor as a whole had not been very successful in the past, attention focused on small groups of poor persons—essentially those who could be reached by a single organization. That focus did much to eliminate the kind of overcommitment to physical construction common in domestic programs.

The prime assumption in the strategy is that social organization is the key to improvement in the human condition. In effect, organization is treated as the major form of social capital, a generalized tool that can be used for a range of purposes. The basic element in the program is a strong local organization, working in the interests of the poor—usually a cooperative. Once established and stabilized, such primary organizations are expected to grow both horizontally and vertically. That is, they will take on additional functions, serve larger numbers of people, and link one primary group to other similar groups through networks of varying size formed to pursue common purposes.

Three aspects of social organization are particularly important for assessing development potential. First, the organizations' purposes must coincide with the needs of their beneficiaries. Those needs are provided for by four basic types of organizational activity, often within a single organization. First, the organization may seek to increase the available supply of goods and services by expansion of production; or by bringing additional services to the local community. Second, it may bring about institutional changes beneficial to the poor. Laws relating to land tenure may be changed, for example. Third, it may increase the relevant knowledge supply. Fourth, it may attempt to foster human development by techniques that range from specialized technical training through what is sometimes called "consciousness raising," that is, making the individual aware of the kind of world in which he or she lives.

The second critical dimension of any development organization is its relation to its beneficiaries. Obviously, the organization must have a genuine commitment to helping the poor. Its procedures should be flexible and humane; its benefits should be fairly distributed. But the IAF strategy goes well beyond these essentials to treating the organization as an opportunity system for the beneficiaries, an opportunity to experience and learn. IAF asks routinely if the organization will: (1) provide the beneficiaries with access to information, resources, and experiences that serve to expand individual potential to the maximum; (2) elevate the public status and private esteem of those involved; and (3) involve itself in activities that go beyond strictly economic matters to include human capitalization through participation in organizational affairs and appropriate education or training. For successful development there must be evidence of efforts at self-help, a clear commitment to the organization, and some element of risk or potential loss for participants. At the same time, the organization must be accountable to its beneficiaries, and machinery for enforcing accountability should be established.

The third essential factor is a sound relationship between funding agency and grantee. The alternative strategy requires a full and responsible partnership in pursuit of mutually agreed purposes, an open relation based on mutual trust. In practice, that kind of partnership entails far more responsibility for the grantee than is usual in foreign assistance. IAF, in effect, provides resources, freedom of action, and moral support. The grantee must supply purposes, staffing, planning, administration, accounting, and needed policy changes. Although IAF may prod and urge (and it must fulfill its statutory requirements relating to finances), it does not interfere, even though the project may fail for lack of intervention. In one sense, this approach is simply a recognition of the superior position of the grantee. No funding agency can maintain constant surveillance over every project. The grantee must be trusted; the IAF strategy simply maximizes that trust. Since learning from experience is a major element in the IAF mandate, such arrangements involve some measure of risk, and some of the weak spots in the IAF records are due to grantee failure.

Responsiveness to Latin American initiative. For the most part, IAF supports efforts at self-help, thus avoiding the paternalism implicit in the traditional strategy. The initiative and requisite skills have emerged within the local community. Local initiative has not confined itself to local affairs. IAF regularly receives proposals to expand networks of organizations to large regions or entire nations. Efforts at international organization are already under way. What is perhaps most important about responsiveness, however, is the implied commitment to assist the poor to seek goals of their own choosing by means of their own device, or, put another way, acceptance of the belief that human priorities are best justified by human willingness to pursue them given the opportunity.

Summarizing, the alternative development strategy emerging from IAF experience is grounded in localism, in a commitment to bottom up development without central planning or direction. The initial focus, analytically at least, is a local population and its primary organization. Such organizations provide the instrument needed to improve conditions of life for the poor and serve as potential anchor points in networks that can extend such benefits to wider populations. There are good reasons, historical, theoretical, and practical, for advocating localism in this form. In the highly volatile political conditions in Latin America it is particularly appropriate to create islands of local strength that can be linked together later if only because there is no way to attack the poverty population as a whole. But the strategy has wider applicability, even in developed society. Localism provides a way of circumventing the principal weakness in centralized administration—the need to make decisions at an administrative level where the consequences of alternative actions cannot be determined.

SOME OPERATIONAL LESSONS

The key element in the alternative strategy is an indigeneous local organization, dedicated to assisting the target population, and suitably related to its sponsors and beneficiaries. Some of these organizations were created by the poor for their own purposes; most were created by an external agent—in Latin America, most often by the church. Most of the organizations that IAF has supported are primary, located in a specific area, and serving a particular population directly. Some are elements in a network of local or regional organizations. Most of the lessons that have been learned working with primary organizations seem to apply equally well to networks.

PRIMARY ORGANIZATIONS

The primary organizations are best distinguished by the types of poor populations they serve. Some have concentrated on groups of isolated or disengaged peoples with few links to existing society; most have worked with rural peasants, desperately poor but with some access to land; a much smaller number have worked with poor urban dwellers. The overall strategy applies in each case, but details vary considerably. The tripartite division is particularly useful for keeping funding agency expectations reasonably well aligned with group potential.

Working with the isolated rural poor in Latin America is a slow and difficult process with few parallels in developed society. These unfortunates must struggle to maintain a bare existence on the fringes of society. Many are unskilled even in farming and do not own such rudimentary tools as hoes or shovels. For this group, the impetus to action must come from outside, and a substantial amount of training is required for even the most rudimentary types of participation in organized activity. The rate of progress is slow. Agencies that fund organizations working with such isolated peoples can expect to maintain contact with the organization for some years and self-reliance is problematic even after substantial periods of time.

The bulk of IAF experience relates to peasants, the most numerous element of Latin American poor. These peasants are attached to the land and to the market system. Their two most serious problems are usually the legal relation to the land and exploitation through the local market. The primacy of economic concerns is striking. Any program that neglects economics entirely will fail, but economic considerations cannot be absolute. Organized peasants will readily forego significant economic benefits to obtain psychic, cultural, or personal satisfactions. Given a choice, the rural poor seem not to choose either to reject economic considerations or abandon themselves entirely to economic pursuits.

IAF experience with urban programs aimed at the urban poor has been limited and the greater complexity of urban poverty makes learning more difficult. As in the United States, urban poverty in Latin America is a function of a range of factors and poverty is not alleviated by supplying a single facility such as housing, education, health care, or even all these things. Ultimately, the condition of life of the urban poor depends on employment and that, in turn, depends on factors that are extremely hard to control, particularly with very modest resources.

Some of the results of IAF experience in urban areas can, however, be generalized. For example, until the prevailing myths relating to the needs and capacities of the poor have been exploded, it is almost impossible for organizations seeking to assist them to obtain support locally. In Latin America, and one suspects in the United States as well, the role that women can and do play in development is widely misunderstood and distorted. Second, assistance programs need to be suited to the available free time of the poor. The urban poor do not, as is often supposed, have a great deal of free time. In most families, several members are either employed or seeking employment and therefore largely occupied. For a poverty household, generating a livelihood is a time-consuming task. Finally, experience suggests that urban programs should be concentrated in a single neighborhood. That allows fuller access to the total population and makes for a more efficient use of resources. Production units should be small, flexible, and suited to the capacities and resources of the residents. One tactic is to treat the neighborhood as an economic trading unit in a complex network. That emphasizes the importance of import substitution. If needed goods can be self-produced or obtained through direct exchange, that reduces the pressure on the very small and difficult to obtain supply of cash available to the poor urban family.

AN URBAN EXAMPLE FROM MEXICO

One of the more fruitful projects funded by IAF is located in a small neighborhood on the outskirts of one of the provincial capitals of Mexico. The population consists primarily of migrants from rural areas (about 60% of the total) and migrants from other parts of the city. The people are very poor, uneducated, unskilled, and in some cases even unfamiliar with a cash economy and commercial employment. Unemployment is common; much of the employment is marginal and temporary. Few earn more than the absolute minimum wage, and many, particularly women, work for much less. The settlement most resembles a refugee camp both physically and socially. Shelters are improvised from whatever materials are available at whatever location can be found. The area is a melange of paths, boulders, buildings, garbage heaps, and cactus clusters, dusty in summer and muddy in winter. Visually, it is a most unpromising area for an assistance program grounded in self-help and aimed at self-sufficiency.

The driving force in the program has been a small group of Mexican educators who organized in 1970 to study the life-style of the residents. In 1972, some members of the organization took up residence in the area and began looking for ways to assist the population rather than study them. Out of that effort has come one of the more compelling experiments in urban development among the very poor to be found in Latin America.

The "program" evolved through a number of stages. Initially, as befits educators, the group concentrated on workshop courses and discussion groups, seeking to increase awareness of urgent local problems. That approach was dropped when they found that discussion of vague goals simply produced ideological divisions and argument. They did, however, manage to identify the three problems considered most significant by the residents: health, housing, and the lack of such public services as garbage disposal and transportation. A work group was formed to deal with each problem area and set the task of finding ways to improve conditions *that could be implemented locally.*

Very slowly, the program took root in the neighborhood. To foster housing improvement, an association was formed in 1976 and a shop established for making building blocks and other construction materials. The shop supplied raw materials at cost, and small amounts of credit were available to the needy. The basic principle was to allow residents to use the shop to create material that could be used directly or sold (in effect, to produce import substitutes). Concurrently, experiments were carried out using other materials for housing construction, producing solar ovens for cooking, and improving health in children. A range of education courses also was offered, but the best device proved to be the weekly work meetings at which members of the association learned how to balance the books and make decisions on loan repayments. Eventually, the enterprise will be turned over to the local residents and run as a cooperative, but the organizers feel the impetus to change must come from the residents.

The shop and credit facilities have been used extensively and responsibly. Over 140 families took part in block construction in the first three years, producing more than 170,000 blocks and various other materials. Sixty-five families borrowed from the loan fund; virtually all of them repaid the loans. In 1978, seventy-three families were interviewed to learn what use had been made of the facility. About one-third had added a new room to their dwellings; some had added more than one room; thirteen had accumulated a supply of blocks and planned to begin construction soon. Another twenty families had made changes in their dwellings amounting to less than a full room of new construction. Eight had sold all of their production. The value of the production is hard to calculate, but one family built a four-room house, much sturdier than the usual dwelling, for less than $300 in cash.

The significance of women's roles in such activities is striking. More than 40% of all work performed in the shop was carried out by women alone and another 25% was performed by women working with men in the family. The psychic benefits of the activity were equally positive: male reaction to female participation was uniformly favorable, and the women interviewed were genuinely grateful for the opportunity to make productive use of time they felt would otherwise have been wasted.

The community health program, though simple and inexpensive, proved spectacularly successful. It has been adopted as a model by one Mexican government health agency and studied by several others. It was created by an organization of women, using a technique developed by Dr. David Morley and known as the "Under Fives" clinic. The basic approach is ultra-simple: the mother maintains an ongoing weight chart for each of her children. Deviations from normal weight increase are used as an early warning signal, and the child is taken to the medical center. The effectiveness of the program is really surprising. To the middle of 1978, some 500 children had been monitored by the health center and more than 300 mothers remained active in the program. The measured infant mortality rate in the sample, which included all participants in 1977, was zero per 1,000; adjusted for sampling error, it amounted to under 35 per 1,000. The rate for the control group, consisting of mothers who lived in the area but did not take part in the program was nearly 110 per 1,000. The official rate for the city was 177 per 1,000.

Efforts to create a consumer cooperative that would supply cheaper food for children failed, but soy flour, an excellent protein source, is sold at cost through the health center. Total operating expenses for 1978 were about $9,000 (U.S.). The income needed to support the health center will be generated locally: garbage recycling, for example, seems to offer a way of earning some of the needed money. The organization seeks self-sufficiency for each of its programs independently; income and expenses for all operations are not lumped.

At IAF request, the working group summarized what had been learned from the first eight years of operations. Five basic points were considered prime:

(1) Unschooled adults are eager to learn concrete ways to meet basic needs and will acquire the skills needed if they are taught

in a practical context. Neither a classroom setting nor a structured course is effective. Unstructured consciousness-raising programs are often counterproductive.

(2) While the technological skills involved in simple programs of this sort might appear rudimentary, a very high level of technical skill is needed to carry out experiments which seek to improve real situations.

(3) Self-sufficiency is not attainable in all cases. When surpluses from production are used to support purely service functions, that is simply philanthropy enforced by some poor on others. Support for service programs should come from the wider community.

(4) While the concept of cooperative action is attractive, particularly to social theoreticians, not all productive ventures need be or should be organized on a cooperative basis. The crucial factor is relevance to the needs, capacities, and resources available in the community.

(5) Even though opportunity is provided for local production, the poor may prefer to do without until they can afford to purchase commercial goods in a regular store.

NETWORKS

A strategic commitment to fostering primary organizations dedicated to improving the lives of the poor extends more or less automatically to the development of networks of such organizations. For isolated and exploited poor populations, networks can increase the supply of necessities in very useful ways. Economies of scale decrease the price of consumer goods and increase the price obtained for local products. Networks are one major device through which the local poor can avoid some of the institutionalized exploitation in society. By spreading burdens and risks across larger populations, they serve to cushion some of the effects of overspecialization or local failures. Assuming that networks are accountable to the poor, they serve to generate mutual support, facilitate exchange of ideas and information, and aggregate influence in ways that help secure additional support from public and private sources.

IAF support for networks increased sharply after 1975, and experience with such organizations is beginning to cumulate. The basic strategy used with primary organizations seems to hold with networks, but there are some differences. Evolutionary patterns may vary. Primary organizations usually do best when they evolve upward from a consenting base. Regional and national organizations have been able to stimulate the creation of successful local groups, particularly if they provide work opportunities or supply specialized services. For example, a cooperative was organized in the capital city of Uruguay to market knitted goods; the central organization very successfully created branch units in various parts of the country to produce the knitted materials.

Probably the most important insight to emerge from IAF experience in networking is the extraordinary importance of finance in large-scale development activities, even those based on self-help and aimed at self-sufficiency. In most cases, IAF efforts to support networks have ended in a search for additional resources in the local money market. There, local organizations of the poor tend to be "redlined" in much the same way that U.S. banks restrict loans in certain areas of the inner city. The networks have therefore found themselves increasingly dependent on external financial expertise, willing or not. For the Latin Americans, finding resources outside the regular banking establishment has become a life or death matter. Development of new financial institutions is difficult because the banking industry tends to be more tightly regulated and controlled than most industries. For that reason, among others, efforts to create worker banks or peasant savings institutions have already begun and seem likely to accelerate. Significantly, the savings have been forthcoming once the institution has been established, suggesting that the poor do have surplus resources but do not trust conventional institutions enough to deposit their savings with them.

APPLYING THE STRATEGY: SOME LIMITS

If the goal is improvement in the lives of the very poor, the bottom-up approach to development is both feasible and effective, but it is not a panacea. A number of limitations and potential sources of failure have already been identified and they bear summary.

SMALL IS NOT ENOUGH

The bottom-up approach to development is not just another way of saying "small is beautiful," though the two are sometimes confused, even within IAF. Staff members have been praised for locating small projects "close to the people," for example, and IAF support for large-scale ventures has been regarded by some with distaste. Yet size guarantees very little. Certainly, the evidence does not support the notion that benefits are more equitably distributed in small organizations or that decision-making is likely to be more democratic. Indeed, small groups may be peculiarly vulnerable to the exercise of special influence.

OVERLOADING

Underdeveloped societies, and poverty areas in general, usually suffer from a severe shortage of skilled and trained manpower. There is therefore an understandable and foreseeable tendency for key personnel in development organizations to be overloaded, accepting too many functions, contributing too much of their time. In the extreme case, one individual becomes so deeply associated with operations that no decision can be made without his or her participation and approval. There is no general solution to overloading of personnel, but a deliberate effort to spread the management burden is essential, even at some serious cost, in order to increase the supply of knowledgable and trained manpower. There will still remain the possibility of serious tension between rate of growth, taking advantage of opportunities, and the dangers of overextension, whatever course of action is decided upon.

DEMYTHOLOGIZING

The importance of myths and mistaken conventional wisdom can hardly be overemphasized. The presumed inadequacy of women, for example, has very seriously damaged a number of Latin American programs. Male administrators have allowed flourishing women's organizations to die, for example, refusing to disburse funds made available for the women by an outside agency. Similarly, the belief that the poor cannot develop the administrative and managerial skills needed to run an organization, or make responsible use of large amounts of resources, is equally widespread. Until such mistaken beliefs are counteracted, access to local resources is likely to remain difficult if not impossible for organizations of the poor.

A somewhat more complex type of mythology has also surfaced during IAF operations. It is best termed the "society-is-the-enemy" approach to arousing the poor to action. It implies hostility to *all* social and political institutions, or to all those not of the proper ideological persuasion. The error lies in the all-inclusive nature of the assumption. It may be reasonable to assume hostility from particular regimes at specific times; in Latin America, that assumption is likely to be valid. But as a permanent attitude, hostility to institutional arrangements in society is self-destructive. No element of any society can prevail over the long run without institutional support; to attack all institutions at all times is to attack society, and that is simply nonsense. The lesson has had to be learned in every society and culture, sometimes at great cost in human suffering. In the long run, society is man's most important artifact and most enduring creation. No program, movement, or organization can survive if it regards the overall society as unremittingly hostile. And no society can survive very long if it contains powerful and utterly irreconcilable elements. The sword cuts both ways.

THE INTRACTABLE PROBLEM

Finally, the bottom-up approach to development brings to the surface, though it cannot solve, the most intractable problem in every society. The assumption that major improvements in the lives of the poor can be brought about through transfer programs alone is simply mistaken. The problem of "development" simply stated is this: society must somehow generate and sustain an adequate level of productive employment among its members. That employment must supply most of the population with the capacity to fulfill its needs by purchase. The economic commonplace that the pie that is divided can be no larger than the pie that is produced simply cannot be ignored. Unfortunately, that sounds precisely like an effort to foster development using the traditional macroeconomic strategy. Not so! Macroeconomic strategies have shown little capacity to deal with the intractable unemployment problem, particularly in the United States. Reliance on "market forces" has been wasteful and exploitative. Perhaps the most fundamental intellectual challenge facing those concerned with the future of the poor is the need to find ways of generating employment that is both productive and rewarding, and associating it with an equitable distribution system. The alternative strategy suggests this be done from the bottom up, particularly in the less-developed nations.

IMPLICATIONS FOR DOMESTIC POLICY

It would be premature to engage in detailed speculation about the implications of the alternative strategy for American domestic assistance programs, but the strategy does appear to have great potential value in domestic affairs and warrants a trial, the more so because the experiment could have a dual function.

An experimental trial in the United States could be undertaken fairly easily in, say, a selected number of small, medium, and large cities. It would have to operate outside the regular bureaucratic machinery to obtain the needed independence but that would actually make it more feasible politically—it would be much more difficult to arrange a trial using resources already allocated to local areas. The test should be arranged to

(a) determine feasibility, and (b) validate the results achieved in Latin America. Selecting a very low-income population as target, and allowing that population to set its own priorities and plan or initiate its own programs, the question to be answered is: will that strategy produce responsible, effective, and self-sustaining activity? A range of tactics may have to be explored: support for profit-making activities that could provide needed employment for the target population, to take one example.

The second aspect of the experiment relates to the most effective way of developing the research capacity needed for making policy. IAF is a first effort to develop that capacity *inside* government rather than in the universities or the private sector—though that was not the intention of the legislation. In some ways, that aspect of the experiment is even more important than its substance. For policy-making is at present in a very dangerous state, mainly because the knowledge and information required for efficient action have not been developed, particularly in the social sciences. The information and knowledge that policy-making requires depend on an intimate and continuing awareness of action and consequences not available outside government. The experiment could test a variety of techniques for obtaining or creating the needed research skills, generating essential knowledge, and putting it to good use. Indeed, it would be an ideal location for the enterprise.

REFERENCES

LEVEN, C. L., J. T. LITTLE, H. O. NOURSE, and R. B. READ (1976) Neighborhood Change: Lessons in the Dynamics of Urban Decay. New York: Praeger.

MEEHAN, E. J. (1975) Public Housing Policy: Convention Versus Reality. New Brunswick, NJ: Center for Urban Policy Research, Rutgers University.

——— (1979a) In Partnership with People: An Alternative Development Strategy. Rosslyn, VA: Inter-American Foundation and U.S. Government Printing Office.

——— (1979b) The Quality of Federal Policy-Making: Programmed Failure in Public Housing. Columbia, MO: Univ. of Missouri Press.

NOURSE, H. O. and D. PHARES (1974) "Socioeconomic transition and housing values: a comparative analysis of urban neighborhoods," pp. 183-208 in G. Gappert and H. M. Rose (eds.) The Social Economy of Cities. Beverly Hills: Sage Publications.

PHARES, D. [ed.] (1977) A Decent Home and Environment: Housing Urban America. Cambridge, MA: Ballinger.

THIESENHUSEN, W. C. (1978) "Reaching the rural poor and the poorest: a goal unmet," pp. 159-182 in H. Newby (ed.) International Perspectives in Rural Sociology. New York: John Wiley.

Appendix: Glossary of Abbreviations

ACIR: Advisory Commission on Intergovernmental Relations
ARFA: Anti-Recession Fiscal Assistance
CBD: Central Business District
CETA: Comprehensive Employment and Training Act
CDBG: Community Development Block Grant
CPD: (Office of) Community Planning and Development, HUD
EDA: Economic Development Administration
FHA: Federal Housing Administration
FHLBB: Federal Home Loan Bank Board
FNMA: Federal National Mortgage Association
FY: Fiscal Year
GAO: General Accounting Office
GIA: General Improvement Area (Britain)
GNMA: Government National Mortgage Association
GRS: General Revenue Sharing
HAA: Housing Action Area (Britain)
HAP: Housing Assistance Plan
HEW: Department of Health, Education and Welfare
HFDAs: Housing Finance and Development Agencies
HUD: Department of Housing and Urban Development
LPW: Local Public Works (Program)
NASPAA: National Association of Schools of Public Affairs and Administration
NHRA: National Housing Rehabilitation Association
NHS: Neighborhood Housing Services
NSA: Neighborhood Strategy Area
OAHD: Office of Assisted Housing Development, HUD
OMB: Office of Management and Budget
PD&R: (Office of) Policy Development and Research, HUD
PN: Priorty Neighborhood (Britain)
SMSA: Standard Metropolitan Statistical Area
UDAG: Urban Development Action Grant

ULI: Urban Land Institute
URBANK: Urban Development Bank (now National Development Bank)
URPG: Urban and Regional Policy Group
VA: Veterans Administration

The Contributors

MARTIN D. ABRAVANEL is the Director of the Division of Policy Studies in HUD's Office of the Assistant Secretary for Policy Development and Research, where he carries out research on housing and community development policies and programs. His other research and publications explore various aspects of the relationship between citizenry and government. He holds a Ph.D. in political science from the University of Wisconsin.

JAMES T. BARRY is a consultant living in Washington, DC. He served on the staff of the National Commission on Neighborhoods from October of 1978 through the conclusion of the Commission's work, first as Deputy Director and then as Director of Research. Barry received a doctorate from the Center for Policy Studies of the State University of New York at Buffalo.

YONG HYO CHO is Professor of Political Science and Head of the Department of Urban Studies at the University of Akron. His research interests are in public policy analysis, particularly in the areas of fiscal relations and urban affairs. During 1977-1978, he served as a Faculty Fellow of the National Association of Schools of Public Affairs and Administration (NASPAA) at HUD. He has authored or coauthored several books, monographs, and research reports including *Determinants of Policy Outcomes in the American States* (1973), *Public Policy and Urban Crime* (1974), and *Measuring the Effects of Legislative Reapportionment in the American States* (1976). Professor Cho is a member of the Editorial Board of *Public Administration Review.*

BILIANA CICIN-SAIN is an Associate Professor of Political Science at the University of California at Santa Barbara where she directs the Marine Policy Program. As a NASPAA Faculty Fellow in 1977-1978 at HUD, she participated in the development of President Carter's national urban policy. Her major publications include a book on the impact of the War On Poverty in a Mexican-American community, *Politicizing the Poor* (1976).

H.W.E. DAVIES has served as Chairman of the School of Planning Studies and Professor of Planning at the University of Reading (England) since July 1977. From 1966 to 1977, he was associated with a British firm of consultants where he was responsible for a study of Liverpool's inner area and for subregional and town plans and related studies in the United Kingdom, Canada, Botswana, and Singapore. Professor Davies was Specialist Advisor on Planning Procedures to the House of Commons Expenditure Committee 1976-1978. He is currently directing the British contribution to the Trinational Inner Cities Project funded by the German Marshall Fund of the United States.

JOHN R. GIST is Associate Professor of Urban Studies at Virginia Polytechnic Institute and State University. His research has focused primarily on budgetary theory, budgeting and appropriations politics, and economic policy. His publications have appeared in *American Journal of Political Science, Journal of Politics,* and in *Sage Professional Papers in American Politics.* From August 1977 to February 1979, he served as an analyst in the Office of Planning and Development at HUD, where he was involved primarily with the Urban Development Action Grant Program. For a year of that time, he served as a NASPAA Fellow. He has also taught at the University of Georgia and Sangamon State University.

RICHARD HULA is an Assistant Professor of Political Economy at the University of Texas at Dallas. His research interests center on government intervention into private markets including the relationship between private home credit and urban development. Other research interests include the management of political controversy and violence.

PAUL K. MANCINI is a Senior Program Analyst in the Division of Policy Studies in HUD's Office of Policy Development and Research. In that capacity, he has been involved in the planning and implementation of numerous major evaluations of housing and community development programs and policies. He has completed all requirements, except the dissertation, for a Ph.D. in political science from the University of Pittsburgh.

EUGENE J. MEEHAN is Professor of Political Science and staff Urban Planner at the University of Missouri-St. Louis. Receiving a degree from London School of Economics, he has taught at the University of Illinois

(Urbana), Brandeis University, and Rutgers University. He is author of *The Quality of Federal Policy-Making: Programmed Failure in Public Housing* (1979), *Public Housing Policy: Myth Versus Reality* (1975), and *Foundations of Political Analysis: Empirical and Normative* (1971). Meehan has also served as consultant for a variety of organizations.

DAVID PURYEAR is currently Associate Professor in the Public Policy Program at Johns Hopkins University. Until recently he served as Director of the Division of Economic Development and Public Finance at HUD. He has also taught at Syracuse University. Puryear's main research areas are in urban public finance and economic development. Among his most recent publications are a coedited special issue of *National Tax Journal* (June 1979), dealing with tax and expenditure limitations, and a coauthored chapter on the urban impact of the Revenue Act of 1978, which will appear in a volume edited by Norman Glickman, *The Urban Impacts of Federal Policies.*

RAYMOND A. ROSENFELD is an Assistant Professor of Political Science at the University of Tulsa, where he specializes in urban policy and public administration. He previously served as an evaluation officer in the Office of Community Planning and Development of HUD, where he directed preparation of HUD's Second Annual Report on the Community Development Block Grant. He is currently working on studies related to the implementation of that program. A portion of his service in Washington was sponsored by a NASPAA fellowship.

DONALD B. ROSENTHAL is Professor and Chairman of the Department of Political Science at the State University of New York at Buffalo. In addition to various publications on local politics in India, he has coauthored *The Politics of Community Conflict* (1969) and articles on general revenue sharing and federal-local relations. His participation in this volume grows out of his experience in 1977-1978 at HUD as a NASPAA Fellow.

JOHN P. ROSS is Director, Division of Economic Development and Public Finance, Office of Policy Development and Research, U.S. Department of Housing and Urban Development. He was formerly associated with the U.S. Advisory Commission on Intergovernmental Relations (ACIR) and coauthored ACIR's study of countercyclical grants, *Federal Stabilization Policy: The Role of State and Local Government.*

HEYWOOD T. SANDERS is an Assistant Professor in the Institute of Government and Public Affairs and the Department of Political Science at the University of Illinois at Champaign-Urbana. He received his Ph.D. from Harvard University. His current research includes an analysis of the federal monitoring of local programs under the Community Development Block Grant program, and an examination of the impact of fiscal distress on local public employment.

ERIC L. STOWE is currently serving as Vice President for Research and Development of the American Chamber of Commerce Executives. During 1977-1978, he was on leave from the University of North Carolina at Charlotte while he served as a NASPAA Fellow at HUD. He worked on the formulation of the 1978 urban policy both there and at the White House Conference on Balanced Growth and Economic Development. His research interests include urban policy and organizational behavior.